MODERN ASPECTS OF
ELECTROCHEMISTRY

No. 4

MODERN ASPECTS OF ELECTROCHEMISTRY

No. 4

Edited by

J. O'M. BOCKRIS

Electrochemistry Laboratory
John Harrison Laboratory of Chemistry
University of Pennsylvania, Philadelphia, Pennsylvania

Suggested U.D.C. *number:* 541.13
Suggested additional number: 621.35

ISBN 978-1-4684-0915-4   ISBN 978-1-4684-0913-0 *(eBook)*
DOI 10.1007/978-1-4684-0913-0

© *Springer Science+Business Media New York*
Originally published by 1996 Plenum Press
Softcover reprint of the hardcover 1st edition 1966

Contents

Contents

Preface

The fourth volume of *Modern Aspects of Electrochemistry* is being prepared at a time of great growth of interest in electrochemistry. The situation can be summarized by saying that the realization is spreading among scientists that electrochemistry represents a broad interdisciplinary field, which has applications to many areas in physics, chemistry, metallurgy, and biology.

Among the reasons for this awakening is the reorientation of what is understood under electrochemistry toward electrodics— "the study of charged interfaces"—with the ionic-solution aspects of electrochemistry being regarded increasingly as aspects of physical chemistry which are helpful auxiliaries to the broad subject of charged interfaces. The pervasiveness of electrochemistry becomes clearer when one recalls that most interfaces carry a charge, or undergo local charge transfers, even though they are not connected with a source of power.

A further reason for the rapid increase in electrochemical studies arises from the technological aspects, in particular in energy conversion and storage, syntheses, extractions, devices, the stability and finishing of surfaces, the treatment of water, etc. The fact that electrodics allows the conversion of chemical to electric energy and the storage of the latter, at the same time producing fresh water as a by-product, presents an aspect of the subject which appears to have far-reaching significance. The intrinsic limitations of atomic reactors to minimum masses too high to provide power sources which can be carried on transports of weight less than hundreds of tons, together with the present growing (and irreversible) world situation with respect to atmospheric pollution (and the CO_2-produced greenhouse effect), gives rise to a close connection between growth in the fraction of our energy which is derived atomically and the probability of the need for electrochemical storage devices and converters in vehicles. It also indicates the urgency of solving the problem of the supply of people trained in modern electrochemistry, where even the present industrial situation (as yet largely unaffected by the future of electrochemical energy conversion and storage) justifies an increase by some seven times in the number of people trained each year (if the ratio of the value of the products of the electrochemical industry to the number of Ph.D's serving it is to equal the average ratio for the whole chemical industry).

The subjects in the present volume have been chosen with the above situation in view. The value of the application of irreversible thermodynamics to electrochemistry is perhaps a controversial one, and that is why a leading expert in this field, Prof. P. Van Rysselberghe, has been invited to express his views in Chapter 1. The mechanism of organic oxidation is a center of modern interest, and if one could solve its problems, direct electrochemical conversion of cheap hydrocarbon fuels to electricity would probably be achieved in a practical and economic way (thereby effecting, because of the greatly increased efficiency of conversion, a technological advance of historic importance). Chapter 2 describes the present state of studies in the mechanism of these important hydrocarbon oxidations, a subject still in its early stages. Professor L. Young and his collaborators present an erudite account of the relatively sophisticated subject of currents in oxide films in Chapter 3.

Lastly, Chapter 4, on the economics of the electrochemical industry in the United States, may cause a number of sharp surprises; for example, the classical electrochemical industry of 1958 and 1963 already accounted for more than 10% of the entire chemical industry, including petroleum products.

J. O'M. BOCKRIS

Philadelphia, September 1966

1

Some Aspects of the Thermodynamic Structure of Electrochemistry

Pierre Van Rysselberghe

I. INTRODUCTION

Except for the use of classical thermodynamics in the treatment of galvanic cells at zero current (including thermogalvanic cells, for which some aspects of the thermodynamics of irreversible processes have to be brought in—a topic not treated here—see Agar's recent review[1] where a complete bibliography will be found) and for the introduction of thermodynamic quantities pertaining to the formation of activated complexes from reactants in electrode processes, little systematic use of thermodynamics has been made in the various areas of electrochemistry concerned with nonzero currents.

By "thermodynamics" we mean the integrated discipline resulting from the incorporation of the thermodynamics of irreversible processes into classical thermodynamics. This discipline, contrary to what is often stated and believed, is not restricted to the range of validity of linear relations between fluxes and forces but to the range of validity of the Gibbs entropy balance equation, a much wider range than that of the linear relations holding near equilibrium. This situation makes thermodynamics the main foundation for the study of electrochemical systems in which irreversible processes take place and, moreover, it gives thermodynamics priority over kinetic considerations in the very field of electrode kinetics. It should be an entirely general line of attack in the study of all types of systems and processes to begin with a thermodynamic analysis and to exhaust the possibilities of

1

thermodynamic reasoning before introducing models and assumptions of a mechanistic or molecular nature, as, e.g., the premature introduction of all the details of an assumed picture of the electrochemical double layer.

In our 1955 monograph on electrochemical affinity,[2] we gave a thermodynamic treatment of cells which are the seats of electric currents, and presented an exploratory thermodynamic approach to electrode kinetics. In 1949,[3] we had already advanced the idea which has remained the basic one in all our subsequent theoretical work in electrode kinetics, namely, that the reacting species in the rate-determining step of the mechanism of an electrode process affect the rate through their electrochemical potentials at the reacting sites. In other words, the arbitrary separation of chemical and electric terms in the electrochemical potentials of the reacting species, followed by separate treatment of these terms, through, e.g., the *a priori* introduction of transfer coefficients in certain electric terms—a procedure implicitly followed in much of current electrode kinetics—cannot be allowed. The "integrity" of electrochemical potentials cannot be violated. In 1958, in a study of steady-state mechanisms of chemical reactions,[4] we introduced a general pattern for reaction velocities of elementary reactions (such as steps of mechanisms) in terms of chemical potentials and affinities (forward, reverse, net) to which we gave the name Marcelin–De Donder[5,6] or MD formula as in the early literature of the Brussels school of thermodynamics where, however, its limitation to elementary reactions was not sufficiently emphasized. In a series of articles[7–12] we have applied the MD formula to electrode processes, the chemical potentials and affinities now being replaced by their electrochemical counterparts. The validity of the method is evident and a coherent theory of electrode kinetics can thereby be developed, as we hope the present review of the subject will demonstrate. Let us note in passing that this MD type of electrode kinetics is in no way incompatible with the absolute-rate theory extended to electrode processes. On the contrary, the connection between these two approaches is much stricter and more clarifying than that between current electrode kinetics and the absolute-rate theory, as we shall show.

In our 1963 monograph on the thermodynamics of irreversible processes,[13] we gave a condensed treatment of galvanic cells, which we shall examine again here in some of its fundamental aspects. We shall also examine the essential points of the thermodynamic approach to general chemical kinetics. This monograph[13] also includes the application of the thermodynamics of irreversible processes similarly treated by other authors and which we shall not consider here: the problem of electric potential gradients in liquid junctions.

The scope of the present review consists of two main parts: (a) The essentials of the thermodynamic treatment of galvanic cells and electrodes out of equilibrium; (b) A detailed account of the thermodynamic approach to electrode kinetics on the basis of the MD method, including a comparison with current electrode kinetics.

The reader who may be somewhat unfamiliar with the thermodynamics of irreversible processes, of which the chemical and electrochemical aspects are of most concern here, will find help, we hope, in our three monographs,[2,13,14] in the two monographs of Prigogine,[15,16] and in the treatise by Prigogine and Defay[17] (in which electrochemistry is not touched, but where the essentials of chemical thermodynamics are clearly presented in terms of affinities and chemical potentials).

The current situation in electrode kinetics is thoroughly presented in Vetter's book.[18] Useful but rather specialized reviews are the recent ones by Parsons[19] and Frumkin.[20,21] Gierst's 1958 thesis[22] and his many valuable later publications show how far an essentially nonthermodynamic point of view can penetrate into the intricacies of electrode kinetics; Hurwitz's 1963 thesis[23] adds to that of Gierst a certain amount of correlation with a thermodynamic leading thread which, however, retains the arbitrary separation of chemical and electric terms in the electrochemical potentials and the a priori introduction of transfer coefficients.

A fundamental aspect of our MD treatment is its acceptance or assumption of the existence, in each surface parallel to the electrode, of local thermodynamic functions and variables and of thermodynamic properties for all species whose individual molecules or ions are centered in that surface. This assumption is equivalent to the statement that we consider only systems for which the Gibbs entropy-balance equation holds locally. In some regions of the interphase between the bulk of the metal and the bulk of the solution, some thermodynamic quantities may exhibit discontinuities.

In a long series of publications[24] Piontelli has developed ideas in close agreement with ours in many respects (fundamental role of electrochemical potentials in electrode kinetics; similar approach to nonequilibrium electrochemistry on the basis of nonequilibrium thermodynamics; presentation of the concepts of electric, chemical, and electrochemical tensions, a subject in which he may have anticipated its first formal presentation in a CITCE report[25]). However, up to now Piontelli has not developed an explicit formulation of electrochemical kinetics in terms of electrochemical potentials and affinities. He has expressed to us his essential agreement with the MD approach. During the last few years other electrochemists have indicated a similar agreement. Among these there are some who appear to be of the opinion that the MD method is only another way of attacking the problems of electrode kinetics, and that it must lead to the same results as the current methods (largely based upon the acceptance of Tafel-like expressions). This is, however, not the case. Significant differences arise, and the greater validity should certainly be assigned to the results obtained from the reasoning which is in greater contact with thermodynamics.

Our presentation will not include any numerical handling of experimental data, our purpose being only to provide a new set of theoretical tools for the unraveling of the mechanisms of electrode processes which convinced readers will then best be able to use in the interpretation of their own experimental work.

II. GALVANIC CELLS AT AND OUT OF
ELECTROCHEMICAL EQUILIBRIUM

1. First and Second Laws of Thermodynamics Applied to
Galvanic Cells

The galvanic cell under study is represented by a diagram, e.g.,

$$\text{Lead } A/\text{Zn}/\text{Zn}^{2+}/\text{Cu}^{2+}/\text{Cu}/\text{Lead } B \qquad (1)$$

where the leads A and B are of the same metal. The inner electric potentials (see CITCE's report on *Electrochemical Nomenclature and Definitions*[26]) of A and B are φ^A and φ^B. We assume that the electric potential difference at the liquid junction is negligible or has been taken into account.

Let us consider an infinitesimal interval of time dt during which a stationary current I (positive or negative) passes through the cell from left to right. The charge $I\,dt$ is thus transported through the cell from electric potential φ^A to electric potential φ^B, while the charge $-I\,dt$ is transported from φ^A to φ^B in the external circuit. The cell does the electric work $I(\varphi^A - \varphi^B)\,dt$ on its surroundings (through the external circuit) at the expense of its internal energy E. Since the heat received by the system (i.e., the cell with its leads, excluding the external circuit) is dQ during dt and the pressure work done by the system is $p\,dV$ we have the following energy balance relation:

$$d\mathbf{E} = dQ - p\,dV - I(\varphi^B - \varphi^A)\,dt \tag{2}$$

which is the expression of the first law of thermodynamics, or the principle of conservation of energy. Equation (2) applies whether the cell reaction occurs in its spontaneous direction [$Zn + Cu^{2+} \rightarrow Zn^{2+} + Cu$, with $I > 0$ in the case of our example (1)] or in its nonspontaneous direction ($Zn^{2+} + Cu \rightarrow Zn + Cu^{2+}$, with $I < 0$).

The second law of thermodynamics states in general that

$$dQ' = T\,dS - dQ \geqslant 0 \tag{3}$$

where S is the entropy of the system, T is the absolute temperature, and dQ' is the so-called Clausius uncompensated heat. The inequality corresponds to irreversible behavior, and the equality corresponds to reversible behavior of the system. Between dQ' and the entropy production d_iS we have the relation

$$dQ' = T\,d_iS \tag{4}$$

and we thus have, calling

$$d_eS = \frac{dQ}{T} \tag{5}$$

the external contribution to dS,

$$T\,dS = T(d_eS + d_iS) \tag{6}$$

and

$$dQ = T\,dS - T\,d_iS \tag{7}$$

Introducing this expression for dQ into (2) we have, for the case of a galvanic cell, the following combined statement of the first and second laws of thermodynamics:

$$dE = T\,dS - p\,dV - I(\varphi^B - \varphi^A)\,dt - dQ' \tag{7a}$$

or

$$d\mathbf{E} = T\,dS - p\,dV - I(\varphi^B - \varphi^A)\,dt - T\,d_iS \tag{7b}$$

2. Reversibility, Short-Circuit, and Intermediate Irreversibility

We shall consider three modes of functioning of the cell which correspond to the same changes of the state functions $\mathbf{E}$, S, and V at the same p and T and for the same amount of charge $dq = I\,dt$ being transferred from left to right (but with I and dt not separately constant).

The first mode of functioning of the cell is that of reversibility, the current I being extremely small, and $dQ' = T\,d_iS$ being zero, in accordance with the second law. We then have

$$d\mathbf{E} = T\,dS - p\,dV - (\varphi^B - \varphi^A)_{\text{rev}}\,dq \tag{8}$$

The subscript rev indicates that $\varphi^B - \varphi^A$ now has a value corresponding to this reversible behavior which will be exactly exhibited by the cell when $I = 0$. Each of the two electrodes is then in a state of electrochemical equilibrium. There can be nonequilibrium states of electrodes with $I = 0$, but we shall not discuss such cases here.

The second mode of functioning of the cell is that of short-circuit, for which $\varphi^A = \varphi^B$. We then have

$$d\mathbf{E} = T\,dS - p\,dV - T\,d_iS_{\text{sc}} \tag{9}$$

The subscript sc indicates that we are dealing here with the short-circuit value of d_iS. In this case the cell is the seat of an ordinary chemical reaction of affinity A progressing by $d\xi$ during the time dt required for the short-circuit current to transfer the charge dq. We have

$$d\mathbf{E} = T\,dS - p\,dV - A\,d\xi \tag{10}$$

Identifying (1), (2), and (3), we find

$$(\varphi^B - \varphi^A)_{\text{rev}}\,dq = T\,d_iS_{\text{sc}} = A\,d\xi \tag{11}$$

Let z represent the number of positive faradays F carried from left to right in the cell diagram when the reaction, as written, progresses by $\Delta\xi = 1$. We then have the following situation for A and z:

With the diagram Zn/... /Cu chosen above, we have

$A > 0$ and $z = +2$ for the reaction $Zn + Cu^{2+} \rightarrow Zn^{2+} + Cu$

$A < 0$ and $z = -2$ for the reaction $Zn^{2+} + Cu \rightarrow Zn + Cu^{2+}$

With the diagram Cu/... /Zn, we would have

$A > 0$ and $z = -2$ for the reaction $Zn + Cu^{2+} \rightarrow Zn^{2+} + Cu$

$A < 0$ and $z = +2$ for the reaction $Zn^{2+} + Cu \rightarrow Zn + Cu^{2+}$

For a given cell diagram, the ratio A/z is invariant. We easily verify that

$$I = zF \frac{d\xi}{dt} \qquad \text{and} \qquad dq = zF \, d\xi \qquad (12)$$

and we obtain from (11),

$$(\varphi^B - \varphi^A)_{\text{rev}} = \frac{T}{zF} \frac{d_i S_{sc}}{d\xi} = \frac{A}{zF} \qquad (13)$$

We have just seen that, for a given cell diagram, A/zF is invariant in magnitude and sign with respect to a reversal in the writing of the cell reaction. With the diagram Zn/... /Cu, we have $A/zF > 0$; with the diagram Cu/... /Zn, we have $A/zF < 0$.

The quantity A/zF is the *electromotive force* (or *chemical tension*) of the cell. It is measured by the electric potential difference at zero current (or open circuit), i.e., $(\varphi^B - \varphi^A)_{\text{rev}}$, electrochemical equilibrium being established at each electrode. We shall write (13) as follows:

$$(\varphi^A - \varphi^B)_{\text{rev}} + \frac{A}{zF} = 0 \qquad (14)$$

The quantity $(\varphi^A - \varphi^B)_{\text{rev}}$, i.e., the difference of inner electric potential between the leads (taken from left to right), is the *reversible electric tension* of the cell, which we shall represent by U_{rev}, whereas, as seen above, A/zF is the *chemical tension* or *e.m.f.* of the cell, which we shall represent by E. We have

$$U_{\text{rev}} + E = 0 \qquad (15)$$

This relation is actually an equilibrium condition between an electric force or tension U_{rev} and a chemical force or tension E. In the case of the Daniell cell (1) with diagram $Zn/\ldots/Cu$ and solutions at unit activity of Zn^{2+} and Cu^{2+} ions, respectively, we have $U_{rev}^0 = -1.1$ V and $E^0 = +1.1$ V; with the diagram $Cu/\ldots$ $/Zn$ we would have $U_{rev}^0 = +1.1$ V and $E^0 = -1.1$ V.

The third mode of functioning of the cell which we shall consider is that of an irreversible transport of charge resulting from the irreversible occurrence of the cell reaction, with the difference $(\varphi^A - \varphi^B)$ different from $(\varphi^A - \varphi^B)_{rev}$ and from zero.

Identifying (9) and (10), and taking (12) into account, we find

$$I(\varphi^A - \varphi^B)\,dt - T\,d_iS = -A\,d\xi \tag{16}$$

or

$$\varphi^A - \varphi^B + \frac{A}{zF} = \frac{T}{zF}\frac{d_iS}{d\xi} \tag{17}$$

We now have an irreversible electric tension $U = \varphi^A - \varphi^B$ different from U_{rev}, but the e.m.f. is still equal to A/zF as long as the physical chemical state (T, p, composition of the bulk phases) remains unchanged. We may write (17) as follows:

$$U + E = \frac{T}{zF}\frac{d_iS}{d\xi} \tag{18}$$

Since d_iS must always be positive, the sign of $U + E$ is determined by that of $z\,d\xi$.

For the diagram $Zn/\ldots/Cu$, we have $E^0 = +1.1$ V. When the reaction is written in its spontaneous direction ($Zn + Cu^{2+} \rightarrow Zn^{2+} + Cu$) and actually occurs in that direction, we have $z = +2$ and $d\xi > 0$. Hence,

$$U > -E^0 \quad\text{or}\quad U > -1.1\text{ V} \quad\text{or}\quad |U| < +1.1\text{ V} \tag{19}$$

The irreversibility of the functioning of the cell makes the absolute value of its electric tension smaller than its chemical tension or e.m.f. If the reaction is still written in its spontaneous direction but actually occurs in the opposite direction, we still have $z = +2$, but now $d\xi < 0$. Hence

$$U < -E^0 \quad\text{or}\quad U < -1.1\text{ V} \quad\text{or}\quad |U| > +1.1\text{ V} \tag{20}$$

Case (19) is that of a galvanic cell producing electric energy; case (20) is that of an electrolytic cell consuming electric energy. The cases of the reaction written in its nonspontaneous direction (with $d\xi < 0$ or > 0) and of the opposite cell diagram with the two ways of writing the reaction (each with $d\xi < 0$ or > 0) are easily treated. We thus have a total of eight possibilities: cell Zn/... /Cu, or cell Cu/... /Zn with reaction $Zn + Cu^{2+} \rightarrow Zn^{2+} + Cu$, or reaction $Zn^{2+} + Cu \rightarrow Zn + Cu^{2+}$ with $d\xi > 0$, or $d\xi < 0$. In the four cases corresponding to production of electric energy we have $|U| < 1.1$ V, and in the four cases corresponding to consumption of electric energy we have $|U| > 1.1$ V.

3. Polarization, Electrochemical Affinity, Electrochemical Tension, and Overtensions

When the cell is traversed by a nonzero current I, it becomes the seat of irreversible phenomena and we have, as shown by (18),

$$U + E = \frac{T}{zF} \frac{d_i S}{d\xi} \neq 0 \qquad (21)$$

At zero current and under reversible conditions we have [see (15)]

$$U_{rev} + E = 0 \qquad (22)$$

Subtracting (22) from (21) we have

$$U - U_{rev} = \frac{T}{zF} \frac{d_i S}{d\xi} \qquad (23)$$

The difference $U - U_{rev}$ is called the *polarization* of the cell. We shall represent it by $\tilde{E}$. We have from (17) and (18)

$$U + E = U + \frac{A}{zF} = \tilde{E} \qquad (24)$$

or

$$zF\tilde{E} = A + zFU = \tilde{A} \qquad (25)$$

We shall call $\tilde{A}$ the *electrochemical affinity* of the cell. It is the sum of the chemical affinity A and of the electric affinity zFU. At equilibrium we have

$$\tilde{A}_{rev} = 0 \qquad \text{or} \qquad A = -zFU_{rev} \qquad (26)$$

Since $E = A/zF$ is the chemical tension of the cell, the polarization

$$\tilde{E} = \frac{\tilde{A}}{zF} \tag{27}$$

can logically be regarded as the *electrochemical tension* of the cell.

Let us decompose U into the sum of electric potential differences at the various contacts in the cell and of the internal ohmic drop RI (R being the internal resistance of the cell) from the solution contiguous to the Zn surface to that contiguous to the Cu surface. We have

$$U = \varphi^A - \varphi^B = (\varphi^A - \varphi^{Zn}) + (\varphi^{Zn} - \varphi^{Zn^{2+}}) + RI$$
$$+ (\varphi^{Cu^{2+}} - \varphi^{Cu}) + (\varphi^{Cu} - \varphi^B) \tag{28}$$

Intermetallic contacts are always regarded as unpolarizable, the Galvani potential differences between metals in contact having then the same values under current as in the absence of current. We then have

$$\tilde{E} = U - U_{rev} = (\varphi^{Zn} - \varphi^{Zn^{2+}}) - (\varphi^{Zn} - \varphi^{Zn^{2+}})_{rev} + RI$$
$$+ (\varphi^{Cu^{2+}} - \varphi^{Cu}) - (\varphi^{Cu^{2+}} - \varphi^{Cu})_{rev} \tag{29}$$

Let us consider the cell reaction as written and as occurring in its spontaneous direction. The Zn electrode is then the anode and we have

$$(\varphi^{Zn} - \varphi^{Zn^{2+}}) - (\varphi^{Zn} - \varphi^{Zn^{2+}})_{rev} = \eta_a > 0 \tag{30}$$

because the entropy production for the anodic process is given by

$$d_iS_a = \frac{1}{T}zF\eta_a\,d\xi > 0 \tag{31}$$

with $z = +2$, $d\xi > 0$, and $I > 0$.

The Cu electrode is then the cathode and we have, taking the differences from Cu to Cu^{2+},

$$(\varphi^{Cu} - \varphi^{Cu^{2+}}) - (\varphi^{Cu} - \varphi^{Cu^{2+}})_{rev} = \eta_c < 0 \tag{32}$$

because the entropy production for the cathodic process is given by

$$d_iS_c = -\frac{1}{T}zF\eta_c\,d\xi > 0 \tag{33}$$

We then have

$$\tilde{E} = \eta_a + RI + |\eta_c| > 0 \tag{34}$$

the entropy production due to the Joule effect being

$$d_iS_J = \frac{1}{T}zFRI\,d\xi = \frac{1}{T}RI^2\,dt \tag{35}$$

If Zn is the cathode and Cu the anode, we easily find that since $z = +2, d\xi < 0$, and $I < 0$,

$$\tilde{E} = -(|\eta_c| + R|I| + \eta_a) < 0 \tag{36}$$

The quantity η_a is called the *anodic overtension* and is always positive. Overtension is synonymous with "overvoltage" and with "overpotential," but we suggest that it is a more preferable appellation. The quantity η_c is called the *cathodic overtension* and is always negative.

The electrochemical affinity defined in (25) is related to the overtensions and the ohmic drop RI as follows:

$$\tilde{A} = zF\tilde{E} = zF\eta_a + zFRI + zF|\eta_c| \tag{37}$$

for the case of production of electric energy ($\tilde{E} > 0$, $\tilde{A}$ has the sign of z), and

$$\tilde{A} = zF\tilde{E} = -zF|\eta_c| - zFR|I| - zF\eta_a \tag{38}$$

for the case of consumption of electric energy ($\tilde{E} < 0$, $\tilde{A}$ has the sign opposite to that of z).

The total electrochemical affinity of the cell can be regarded as the sum of an anodic electrochemical affinity $\tilde{A}_a = \pm zF\eta_a$, a cathodic electrochemical affinity $\tilde{A}_c = \mp zF\eta_c$, and a Joule effect electrochemical affinity $\tilde{A}_J = \pm zFRI$.

The connection between overtensions and electrochemical affinities, which so far has received little attention by electrochemists, is of fundamental importance and affords a most useful channel for the introduction of thermodynamic considerations into electrode kinetics.

The following formulas, directly derived from the above discussion, will often be found useful:

For a cell generating electric energy the cell tension is given by

$$|U| = |E| - (\eta_a + |\eta_c| + R|I|) \tag{39}$$

and the RI-corrected electric tension (or galvanic tension) is

$$|U| + R|I| = |E| - (\eta_a + |\eta_c|) \tag{40}$$

For a cell consuming electric energy the cell tension is given by

$$|U| = |E| + (\eta_a + |\eta_c| + R|I|) \tag{41}$$

and the RI-corrected electric tension (or galvanic tension) is

$$|U| - R|I| = |E| + (\eta_a + |\eta_c|) \tag{42}$$

The quantities

$$\eta_a + |\eta_c| = |E| - |U| - R|I| = |\tilde{E}| - R|I| \tag{43}$$

for a generating cell, and

$$\eta_a + |\eta_c| = |U| - |E| - R|I| = |\tilde{E}| - R|I| \tag{44}$$

for a consuming cell can be regarded as RI-corrected polarizations, whereas $|\tilde{E}|$ is the total polarization or electrochemical tension of the cell in absolute value.

4. Comparison with Textbook Presentations

Our formulas (17), (22), etc., give us

$$U_{\text{rev}} + E = U_{\text{rev}} + \frac{A}{zF} = 0 \tag{45}$$

The affinity A is related to the free-enthalpy change ΔG for one occurrence of the cell reaction as follows:

$$A = -\left(\frac{\partial G}{\partial \xi}\right)_{Tp} \equiv -\Delta G \tag{46}$$

We note that Δ represents the operator $\partial/\partial \xi$ and not a finite difference. Formula (45) can thus be written

$$U_{\text{rev}} - \frac{\Delta G}{zF} = 0 \tag{47}$$

with

$$\Delta G = -zFE \tag{48}$$

This last formula appears to be identical with that found in many textbooks, except that we are giving a sign to z, whereas z is, as a rule, taken as being always positive in the textbooks. The method followed in our presentation requires that z be given a positive sign when the cell reaction, occurring as written, carries a positive charge from left to right in the cell diagram, and given a negative sign in the opposite case. We have seen that our method implies that E is invariant with respect to a reversal in the writing of the cell reaction. If we write (48) as

$$E = -\frac{\Delta G}{zF} \tag{48a}$$

and take the $Zn/\ldots/Cu$ diagram, we see that $E > 0$, $\Delta G < 0$ with $z > 0$ when the reaction is written $Zn + Cu^{2+} \rightarrow Zn^{2+} + Cu$, and $E > 0$, $\Delta G > 0$ with $z < 0$ if the reaction is written $Zn^{2+} + Cu \rightarrow Zn + Cu^{2+}$.

In any case, whether we take an algebraic z or an arithmetic z, formulas (48) or (48a) actually have no electric content. The quantities $\Delta G = -A$ and E are of the same type and correspond to the linear combination of the chemical potentials μ_i of reactants and products of the cell reaction:

$$\Delta G = -A = \sum_i \nu_i \mu_i \tag{49}$$

and

$$E = \frac{A}{zF} = -\frac{\Delta G}{zF} = -\frac{\sum_i \nu_i \mu_i}{zF} \tag{50}$$

the stoichiometric coefficients ν_i being positive for products and negative for reactants.

The electric information results from the fact that, at equilibrium, $U_{rev} = -E$. This leads us to suggest that the textbook formula (48) should be written

$$\Delta G = zFU_{rev} \tag{51}$$

it being understood that $\Delta G = -E/zF$. It is thus made clear that at equilibrium, the chemical tension is balanced by the electric tension.

Formula (48) is of no help when irreversible behavior occurs because E remains constant for a given physical chemical state of the system; on the other hand, the electric tension U is a function of the current I. As already seen in section 3 we have, out of equilibrium,

$$U + E = U + \frac{A}{zF} = U - \frac{\Delta G}{zF} = \tilde{E} \neq 0 \qquad (52)$$

Let us add at this point that the electrochemical affinity $\tilde{A}$ of the cell reaction and the electrochemical tension $\tilde{E}$ may be written under forms similar to (49) and (50):

$$\tilde{A} = A + zFU = - \sum_i v_i \tilde{\mu}_i \qquad (53)$$

$$\tilde{E} = \frac{\tilde{A}}{zF} = - \frac{\sum_i v_i \tilde{\mu}_i}{zF} \qquad (54)$$

where the $\tilde{\mu}_i$'s are the *electrochemical potentials* of reactants and products, which are of the form

$$\tilde{\mu}_i = \mu_i + z_i F \varphi_i \qquad (55)$$

z_i being the charge number of constituent i (positive for a cation, negative for an anion, zero for a neutral species) and φ_i is the inner electric potential of the phase where i is located.

5. Single Electrodes, Their Electric and Chemical Tensions or Potentials

Let us use as the electrode on the right-hand side of the cell diagram a standard hydrogen electrode, and let us keep the Zn/Zn^{2+} electrode on the left-hand side:

$$\text{Lead } A/Zn/Zn^{2+}/H^+/H_2, Pt/\text{Lead } B \qquad (56)$$

The chemical tension or *e.m.f.* of this cell is equal to $(\varphi^B - \varphi^A)_{rev}$, whereas its electric tension is $\varphi^A - \varphi^B$. Writing the cell reaction as $Zn + 2H^+ \rightarrow Zn^{2+} + H_2$, we have $z = +2$ and

$$E = (\varphi^B - \varphi^A)_{rev} = \frac{A_{ox}}{2F} = E_{ox} \qquad (57)$$

where A_{ox} is the *oxidation affinity* of Zn to Zn^{2+} by H^+ at activity 1 becoming H_2 at the pressure of 1 atm. The ratio $A_{ox}/2F = E_{ox}$ is the *oxidation affinity per unit charge* or *oxidation tension* or *oxidation potential*. We thus have, at equilibrium,

$$U_{rev} + E_{ox} = 0 \qquad \text{or} \qquad U_{rev} = -E_{ox} \qquad (58)$$

With current and Zn/Zn^{2+} the anode, we have $U + E_{ox} > 0$; if Zn/Zn^{2+} is the cathode we have $U + E_{ox} < 0$.

In the spirit of the 1953 Stockholm recommendation[27] (tentatively reformulated at Copenhagen in 1964), E_{ox} *may not* be called the electrode potential of the Zn/Zn^{2+} electrode. We have [see (50)]

$$E_{ox} = \frac{1}{2F}(\mu_{Zn} + 2\mu^0_{H^+} - \mu_{Zn^{2+}} - \mu^0_{H_2}) \qquad (59)$$

We may rewrite the equilibrium condition (58) as follows:

$$U_{rev} = -E_{ox} = E_{red} \qquad (60)$$

where

$$E_{red} = \frac{A_{red}}{2F} = \frac{1}{2F}(\mu_{Zn^{2+}} + \mu^0_{H_2} - \mu_{Cu^{2+}} - 2\mu^0_{H^+}) \qquad (61)$$

A_{red} being the *reduction affinity* of Zn^{2+} to Zn by H_2 at 1 atm becoming H^+ at activity 1. The ratio $A_{red}/2F = E_{red}$ is the *reduction affinity per unit charge* or *reduction tension* or *reduction potential*. We see that E_{red} is the chemical tension or e.m.f. of the cell

$$\text{Lead } A/\text{Pt, } H_2/H^+/Zn^{2+}/Zn/\text{Lead } B \qquad (62)$$

In the spirit of the Stockholm recommendation we may consider E_{red} as equal to the electrode potential of the Zn/Zn^{2+} electrode for the case of zero current. Actually the electric tension of cell (56) is always, with and without current, that of the Zn/Zn^{2+} electrode taken from metal to solution relatively to the conventionally selected standard hydrogen electrode.

Summarizing, we have in the typical case of the Zn/Zn^{2+} electrode or couple (at 25°C in water),

Electric tension of the electrode:

$$U = \varphi^A - \varphi^B \qquad \text{in cell (56)}$$

Reversible electric tension of the electrode (standard):

$$U^0_{rev} = (\varphi^A - \varphi^B)^0_{rev} = -0.76 \text{ V} \qquad \text{in cell (56)}$$

Electrode potential of the electrode (standard):

$$E^0 = (\varphi^B - \varphi^A)^0_{rev} = -0.76 \text{ V} \qquad \text{in cell (62)}$$

Oxidation potential (or tension) of the couple (standard):

$$E^0_{ox} = (\varphi^B - \varphi^A)^0_{rev} = +0.76 \text{ V} \qquad \text{in cell (56)}$$

Reduction potential (or tension) of the couple (standard):

$$E^0_{red} = (\varphi^B - \varphi^A)^0_{rev} = -0.76 \text{ V} \qquad \text{in cell (62)}$$

We have suggested[28,29] that the electromotive series be given in textbooks with three columns of data as follows:

Table 1

Standard Electric and Chemical Tensions of Electrodes or Couples

Electrodes or couples	Reversible electric tensions, Volt	Oxidation tensions, Volt	Reduction tensions, Volt
Cu/Cu^{2+}	+0.34	−0.34	+0.34
H_2/H^+	0	0	0
Zn/Zn^{2+}	−0.76	+0.76	−0.76

The study of the behavior of single electrodes is in principle that of cells in which the other electrode is a reversible reference electrode, the standard hydrogen electrode, or some other suitable reversible electrode. It is advisable to consider the current passing from metal to solution at the electrode under study as positive when the electrode is functioning as an anode and, similarly, as negative when the electrode is functioning as a cathode. Actually, the electrode under study is on the left-hand side of the cell diagram, and the current I appearing in the various formulas above has the sign properties which have just been stated. In this manner the entropy production due to an electrode process is always proportional to a product of two factors of the same sign. At a single electrode we have

$$T \frac{d_i S}{dt} = \eta I \geqslant 0 \qquad (63)$$

For anodic functioning $\eta_a > 0$ with $I > 0$; for cathodic functioning $\eta_c < 0$ with $I < 0$. Since any net current I is the resultant of an anodic component I_a and of a cathodic component I_c, we have

$$I = I_a + I_c = I_a - |I_c| \qquad (64)$$

Let us note that the majority of electrochemists use the opposite convention: $\eta_a > 0$ with $I < 0$; $\eta_c < 0$ with $I > 0$. This is, however, thermodynamically illogical because the entropy production is then proportional to $-\eta I$.

Formula (63) can be written as

$$\eta I = (U - U_{rev})I \geqslant 0 \qquad (65)$$

where the U's are electric tensions in cells in which the reference electrode is on the right. When $U > U_{rev}$ we have $I > 0$ and the electrode is an anode; when $U < U_{rev}$ we have $I < 0$ and the electrode is a cathode.

III. THERMODYNAMIC APPROACH TO ELECTRODE KINETICS

1. Proportionality Between Small Currents and Small Overtensions

When U differs slightly from its equilibrium value U_{rev} (and when the electrode is not "indifferent" in the sense defined in the CITCE report[26]), we must logically expect a proportionality between the cause $U - U_{rev} = \eta$ and the effect I as is the case of all physical chemical processes occurring in the vicinities of their equilibrium states. We write

$$I = l\eta \qquad (66)$$

where l is a phenomenological coefficient of the self-influence type. We have $l > 0$.

From (37) we see that the electrochemical affinity $\tilde{A}$ of a cell consists of three contributions due to the anodic and cathodic overtensions and to the ohmic drop. If one of the overtensions is zero, as is the case for the reference electrode of the tensiometric cell, and if an RI correction is made, we can write for a single electrode

$$\tilde{A} = zF\eta \qquad (67)$$

We have z and η both positive (anodic functioning) or z and η both negative (cathodic functioning). From (66) and (67) we get

$$I = \frac{l}{zF} \tilde{A} \qquad (68)$$

The entropy production is given by

$$T \frac{d_i S}{dt} = l\eta^2 = \frac{l}{(zF)^2} \tilde{A}^2 \tag{69}$$

and also by

$$T \frac{d_i S}{dt} = \left(\frac{1}{l}\right) I^2 \tag{70}$$

2. Layers in the Interphase Regions and Corresponding Overtensions

Information about the coefficient l will be obtained as a limiting case of kinetic expressions more general than (66). In order to establish these kinetic expressions we must first examine how the overall overtension η can be broken into portions pertaining to the various regions of the total interphase layer between the bulk of the metal phase and the bulk of the solution phase. These regions have been carefully defined in chapter 6 of the CITCE report.[26] We shall simply list them here, going from right to left, i.e., from solution to metal: the diffusion layer dc, the diffuse layer cb, and the transfer layer ba, the bulk phases being the metal I and the solution II, as shown in the following diagram:

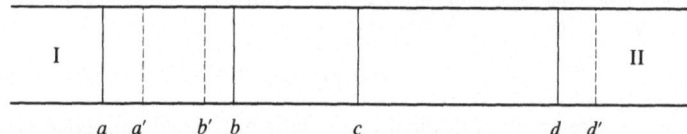

$$a \quad a' \qquad b' \quad b \qquad\qquad c \qquad\qquad\qquad d \quad d'$$

In the absence of current the diffusion layer dc does not exist, the bulk composition of the solution II extending to c where, toward the left, a space charge begins to exhibit itself as a function of the distance to surface b from which, toward the left again, various adsorption equilibria are established and local space charges may also exist. The standard chemical potentials of the various species present in the solution have the same values in cb as in the bulk solution II, whereas the standard chemical potentials of the species present in ba are in general different from their values in II. Layers cb and ba considered together constitute the *electrochemical double layer*.

When current is passing through the interphase, the circumstances which determine the *mass transfer* of the constituents from the bulk solution II toward the electrode or vice versa cause the formation of the diffusion layer dc in which the composition varies from d to c, no significant space charges being created. The properties of layers cb and ba will depart from their values at zero current since the inner electric potential of metal I is then different from its equilibrium, zero current value. We shall conventionally take $\varphi^{II} = 0$ and the overtension is thus $\eta = \varphi^I - \varphi^I_{rev}$. In order to find out how, theoretically at least, η can be broken up into contributions from the three portions of the interphase, we must decompose the total electrochemical affinity of the electrode process into corresponding portions. We shall do this for the simple case of the cathodic reduction of a cation X^{x+} to the metal X, generalization to any type of process being then trivial. The overall reaction is

$$X^{x+}(II) + xe^-(I) \rightarrow X(I) \qquad (71)$$

We have
$$\tilde{A} = zF\eta = -xF\eta \qquad (72)$$

$$I = zFv = -xFv \qquad (73)$$

where v is the reaction velocity $d\xi/dt$. If the electrode process actually takes place in the direction (71), we have $\eta < 0$ and $v > 0$; then, since $x > 0$, $\tilde{A} > 0$ and $I < 0$. If the process takes place in the anodic direction, the reaction being still written as in (71), we have $\eta > 0$ and $v < 0$; then, since $x > 0$, $\tilde{A} < 0$ and $I > 0$. Our procedure here is thus entirely consistent with the discussion in Part I.

In order to avoid dimensional difficulties we shall explicitly introduce the area of the electrode (assumed plane) and write for the current density,

$$i = \frac{I}{s} = -\frac{xFv}{s} \qquad (74)$$

In terms of electrochemical and chemical potentials we have, for (71),

$$\tilde{A} = \tilde{\mu}^{II}_{X^{x+}} + x\tilde{\mu}^I_{e^-} - \mu^I_X \qquad (75)$$

The ion X^{x+}, coming from the bulk of phase II, first traverses the diffusion layer dc and there is a corresponding electrochemical affinity:

$$\tilde{A}^{dc} = \tilde{\mu}^d_{X^{x+}} - \tilde{\mu}^c_{X^{x+}} \qquad (76)$$

The ion then traverses the diffusion layer cb, with

$$\tilde{A}^{cb} = \tilde{\mu}_{X^{x+}}^c - \tilde{\mu}_{X^{x+}}^b \tag{77}$$

In the transfer layer ba the ion is discharged and we have

$$\tilde{A}^{ba} = \tilde{\mu}_{X^{x+}}^b + x\tilde{\mu}_{e^-}^a - \mu_X^a \tag{78}$$

It is easily verified that

$$\tilde{A} = \tilde{A}^{dc} + \tilde{A}^{cb} + \tilde{A}^{ba} \tag{79}$$

We then have corresponding overtensions:

$$\eta^{cd} = -\frac{\tilde{A}^{dc}}{xF} = \frac{\mu_{X^{x+}}^d - \mu_{X^{x+}}^c}{-xF} + \varphi^c - \varphi^d \tag{80}$$

$$\eta^{bc} = -\frac{\tilde{A}^{cb}}{xF} = \frac{\mu_{X^{x+}}^c - \mu_{X^{x+}}^b}{-xF} + \varphi^b - \varphi^c \tag{81}$$

$$\eta^{ab} = -\frac{\tilde{A}^{ba}}{xF} = \frac{\mu_{X^{x+}}^b + x\mu_{e^-}^a - \mu_X^a}{-xF} + \varphi^a - \varphi^b \tag{82}$$

The measured electric tension of the electrode under current is affected by an uncertainty because of the fact that the tip of the Haber–Luggin capillary necessary for its measurement cannot be placed exactly in d, where the composition begins to differ from that in II, but is in d' on the right of d, and the required ohmic correction for the distance dd' cannot be made accurately. In the present review we shall assume that both this correction and the whole mass transfer overtension η^{cd} either have been estimated and corrected for or are negligible. Moreover, we shall in no way concern ourselves with cases where mass transfer is the rate-determining step in the total mechanism of the electrode process. Our discussion will imply that the composition in c is that with respect to which everything occurring on the left of c is referred to. In other words, for our purpose, we shall identify the composition in c with the bulk composition. This is actually the case whenever mass transfer is rapid enough to make some particular step of the discharge process rate-determining. We shall thus be essentially concerned with the *discharge* (or *charge*) *kinetics*, allowing for the possibility that the kinetics of passage from c to b may have an influence on the $i = i(\eta)$ relationship, the corresponding electrochemical affinity A^{cb} being then different from zero.

Using the convenient language adopted by Gierst[22] and Hurwitz[23] we may have to concern ourselves with the "approach kinetics" even when the rate-determining step is part of the discharge process.* Let us note that the process of approach may not terminate exactly in b but may require penetration to b' where the μ^0's are different from their values in cb. Similarly, the reacting electron may be in a state or position a' different from a and there can then also be a problem of "approach kinetics." There can also be an affinity contribution corresponding to the passage of the metal atoms X formed in state a'' to the state a corresponding to the bulk of the metal. The electrochemical affinity which we shall consider pertains to the double layer ca and consists then, in general, of the following contributions:

$$\tilde{A}^{ca} = (\tilde{\mu}^c_{X^{x+}} - \tilde{\mu}^{b'}_{X^{x+}}) + x(\tilde{\mu}^a_{e^-} - \tilde{\mu}^{a'}_{e^-})$$
$$+ (\tilde{\mu}^{b'}_{X^{x+}} + x\tilde{\mu}^{a'}_{e^-} - \mu^{a''}_X) + (\mu^{a''}_X - \mu^a_X) \qquad (83)$$

The first, second, and fourth are approach terms which may, together or separately, be negligible, small, or perhaps large compared with the third term which is that corresponding to discharge. We shall consider all such possibilities and their influences on the kinetics of the overall electrode process.

3. Some Fundamental Considerations on Chemical Kinetics

Let us consider a heterogeneous reaction whose velocity for the area of separation between the two phases is

$$v = \frac{d\xi}{dt} \qquad (84)$$

Per unit area we have

$$\frac{v}{s} = \frac{1}{s} \frac{d\xi}{dt} \qquad (85)$$

the dimensions of the left- and right-hand terms being $(\text{length})^{-2}$ $(\text{time})^{-1}$.

---

*Gierst and Hurwitz speak of "cinétique d'apport" for mass transfer kinetics, of "cinétique d'approche" and of "cinétique de décharge" or "de charge."

The velocity v is the difference between a forward and a reverse velocity:

$$v = \vec{v} - \overleftarrow{v} \qquad (86)$$

In the case of a simple mechanism each of the two ratios, $\vec{v}/s$ and $\overleftarrow{v}/s$, would usually be equal to specific rates $\vec{k}$ and $\overleftarrow{k}$, respectively, multiplied by certain concentrations raised to small integral powers such as 1, 2, or 3. Let us take the simple case of v proportional to a concentration C_M of reactant, and v proportional to a concentration C_N of product N:

$$\frac{v}{s} = \vec{k}C_M - \overleftarrow{k}C_N \qquad (87)$$

Since v/s has the dimensions $(\text{length})^{-2}(\text{time})^{-1}$ we see that $\vec{k}$ and $\overleftarrow{k}$ have the dimensions length $\times$ $(\text{time})^{-1}$ when the concentrations C_M and C_N are volume concentrations. However, the rates may depend on surface concentrations, in which case $\vec{k}$ and $\overleftarrow{k}$ have the dimension $(\text{time})^{-1}$.

We could write C_M and C_N in terms of the corresponding chemical potentials, assuming ideality for the present:

$$C_M = \exp\left(\frac{\mu_M - \mu_M^0}{RT}\right) \qquad C_N = \exp\left(\frac{\mu_N - \mu_N^0}{RT}\right) \qquad (88)$$

with C_M and C_N being now pure numbers. We have

$$\frac{v}{s} = \vec{k} \exp\left(-\frac{\mu_M^0}{RT}\right) \exp\left(\frac{\mu_M}{RT}\right) - \overleftarrow{k} \exp\left(-\frac{\mu_N^0}{RT}\right) \exp\left(\frac{\mu_N}{RT}\right) \qquad (89)$$

the dimensions of $\vec{k}$ and $\overleftarrow{k}$ being then $(\text{length})^{-2}(\text{time})^{-1}$.

When the reaction $M \to N$ is at equilibrium we have $\mu_M = \mu_N$ and it follows that

$$\vec{k} \exp\left(-\frac{\mu_M^0}{RT}\right) = \overleftarrow{k} \exp\left(-\frac{\mu_N^0}{RT}\right) = \frac{1}{\lambda s} \qquad (90)$$

where λ is a function of temperature, and pressure only having the dimension of time. The velocity v is then

$$v = \frac{1}{\lambda}\left[\exp\left(\frac{\mu_M}{RT}\right) - \exp\left(\frac{\mu_N}{RT}\right)\right] \qquad (91)$$

This is the formula which, in its generalized form in terms of forward and reverse affinities of elementary reactions, we have called the *Marcelin–De Donder formula*, in agreement with the literature from the Brussels school:

$$v = \frac{1}{\lambda}\left[\exp\left(\frac{\vec{A}}{RT}\right) - \exp\left(\frac{\overleftarrow{A}}{RT}\right)\right] \tag{92}$$

Under this form, v and $1/\lambda$ are extensive quantities and v/s and $1/\lambda s$ are intensive quantities. Formula (92) is of the same form for homogeneous reactions, v/V and $1/\lambda V$ then being intensive.

Formula (92) covers the following three types of cases (it must be kept in mind that $\vec{A}$, $\overleftarrow{A}$, $A = \vec{A} - \overleftarrow{A}$ pertain to the rate-determining step which may or may not concentrate in itself the overall affinity):

1) The reverse velocity is negligible because A has a large negative value:

$$v = \frac{1}{\lambda}\exp\left(\frac{\vec{A}}{RT}\right) = \vec{v}_e\exp\left(\frac{\vec{A} - \vec{A}_e}{RT}\right) \tag{93}$$

where $\vec{v}_e = \overleftarrow{v}_e$ is the exchange velocity making $v = \vec{v}_e - \overleftarrow{v}_e = 0$ at equilibrium, and $\vec{A}_e = \overleftarrow{A}_e$ is the equilibrium value of either the forward or the reverse affinity.

2) The reverse velocity is not negligible. We have

$$v = \vec{v}_e\left[\exp\left(\frac{\vec{A} - \vec{A}_e}{RT}\right) - \exp\left(\frac{\overleftarrow{A} - \overleftarrow{A}_e}{RT}\right)\right] \tag{94}$$

3) The system is near equilibrium. Both exponents in (94) are small and we have, since $A = \vec{A} - \overleftarrow{A}$,

$$v = \vec{v}_e\,\frac{A}{RT} \tag{95}$$

This last formula is of the form of a proportionality relation between cause A and effect v for small departures from equilibrium where $v = 0$ and $A = 0$. It is thus of the same type as formula (66).

4. Fundamentals of Electrode Kinetics on the Basis of the Marcelin–De Donder Formula

Let us replace A of (95) by the electrochemical affinity $\tilde{A} = zF\eta$. We have

$$v = \vec{v}_e \, \frac{\tilde{A}}{RT} = \vec{v}_e \, \frac{zF\eta}{RT} \tag{96}$$

The current is given by

$$I = zFv = zF\vec{v}_e \, \frac{\tilde{A}}{RT} = I_o \, \frac{z}{|z|} \, \frac{\tilde{A}}{RT} \tag{97}$$

where I_o is the *exchange current* taken as positive and equal to $|z|Fv_e$. We see that l in (66) is given by

$$l = \frac{I_0|z|F}{RT} \tag{98}$$

Let us consider the mechanism of the electrode process as a sequence of elementary reactions all of which, in a steady state, occur at the same velocity, equal to the overall velocity, when stoichiometry is properly taken into account. Let us designate by j the rate-determining step and let us assume that it occurs γ_j times during one occurrence of the overall process. The number γ_j is the *stoichiometric number* of step j. Let us consider the system as being in a state close to electrochemical equilibrium for which it can safely be assumed that the electrochemical affinities of all other steps are equal to zero. We thus have

$$\tilde{A} = \gamma_j \tilde{A}_j \tag{99}$$

Writing (95) more explicitly for step j we have

$$v_j = \vec{v}_{je} \, \frac{\tilde{A}_j}{RT} \tag{100}$$

where $\vec{v}_{je}$ being positive, v_j and $\tilde{A}_j$ have the same sign.
The overall velocity is given by

$$v = \frac{v_j}{\gamma_j} = \vec{v}_{je} \frac{\tilde{A}_j}{\gamma_j RT} \tag{101}$$

and the current density $i = I/s$ by

$$i = |z| \frac{Fv}{s} = \vec{v}_{je} \frac{|z|F}{s\gamma_j} \frac{\tilde{A}_j}{RT} = i_{oj} \frac{\tilde{A}_j}{RT} \tag{102}$$

where i_{oj} is the exchange current density characteristic of step j and is taken positive. We have

$$i_{oj} = \vec{v}_{je} \frac{|z|F}{s\gamma_j} \tag{103}$$

Let us note in passing that the net current i is equal to i_{oj} when $\tilde{A}_j = RT$.

Let us now go back to the case of large departures from electrochemical equilibrium. Formulas (92) and (94) become, in the electrochemical case,

$$i = \frac{zF}{\lambda_j s\gamma_i} \left[\exp\left(\frac{\tilde{\tilde{A}}_j}{RT}\right) - \exp\left(\frac{\tilde{\tilde{A}}_j}{RT}\right) \right] \tag{104}$$

and

$$i = i_o \frac{z}{|z|} \left[\exp\left(\frac{\tilde{\tilde{A}}_j - \tilde{\tilde{A}}_{je}}{RT}\right) - \exp\left(\frac{\tilde{\tilde{A}}_j - \tilde{\tilde{A}}_{je}}{RT}\right) \right] \tag{105}$$

with

$$i_o = \frac{|z|F}{\lambda_j s\gamma_i} \exp\left(\frac{\tilde{\tilde{A}}_{je}}{RT}\right) \tag{106}$$

These last three formulas constitute the foundations of MD electrode kinetics.

5. Particular Case of the Empirically Observed Tafel Behavior

A fairly large number of experimental studies appear to yield data which follow, over appreciable ranges of overtension and current, the so-called Tafel behavior represented by the formula

$$i = i_o \left[\exp\left(\frac{\beta_a F\eta}{RT}\right) - \exp\left(-\frac{\beta_c F\eta}{RT}\right) \right] \tag{107}$$

where the parameters β_a and β_c will be called *overtension factors*; β_a is the anodic and β_c the cathodic overtension factor.

Identification of the Tafel formula (107) with the MD formula (105) gives us the following correspondence between the quantities appearing in these two formulas, the arrows being now replaced by the subscripts a and c:

$$\tilde{A}_{ja} - \tilde{A}_{je} = \beta_a F\eta \qquad \tilde{A}_{jc} - \tilde{A}_{je} = -\beta_c F\eta \qquad (108)$$

Subtracting the second of these relations from the first, remembering that $\tilde{A}_{ja} - \tilde{A}_{jc} = \tilde{A}$ and writing $\tilde{A} = |z|F\eta$ (because $\tilde{A}$ corresponds here to the anodic direction), we obtain

$$\tilde{A}_j = (\beta_a + \beta_c)F\eta = \frac{(\beta_a + \beta_c)\tilde{A}}{|z|}$$
$$= \frac{(\beta_a + \beta_c)\tilde{A}}{x} \qquad (109)$$

where x is the stoichiometric factor of the electron in the overall process.

Let us first consider the case where step j concentrates the total electrochemical affinity of the process, making $\tilde{A}_1, \ldots \tilde{A}_{j-1}$, $\tilde{A}_{j+1}, \ldots$ practically equal to zero. For instance, j could be a charge or discharge step with all approach steps practically at their respective equilibria (see the discussion in section II.2). We then have $\gamma_j \tilde{A}_j = \tilde{A}$ [see (99)], and from (109) we obtain

$$\beta_a + \beta_c = \frac{x}{\gamma_j} \qquad (110)$$

Let us now introduce the parameters α_{aj} and α_{cj} such that

$$\alpha_{aj} = \frac{\beta_a \gamma_j}{x} \qquad \alpha_{cj} = \frac{\beta_c \gamma_j}{x} \qquad (111)$$

with

$$\alpha_{aj} + \alpha_{cj} = 1 \qquad (112)$$

and

$$\alpha_{aj} = \frac{\beta_a}{\beta_a + \beta_c} \qquad \alpha_{cj} = \frac{\beta_c}{\beta_a + \beta_c} \qquad (113)$$

These parameters α_{aj} and α_{cj} are thus derivable from the experimentally obtained overtension factors β_a and β_c. From (108) we find

$$\tilde{A}_{ja} - \tilde{A}_{je} = \alpha_{aj}\tilde{A}_j \qquad \tilde{A}_{jc} - \tilde{A}_{je} = -\alpha_{cj}\tilde{A}_j \qquad (114)$$

The parameters α_{aj} and α_{cj} are the true *transfer coefficients* of the rate-determining step j. The MD method gives a clear meaning to these coefficients.

Let us now consider the case in which step j does not involve the total electrochemical affinity of the overall process but only the constant fraction δ_j of $\tilde{A}$:

$$\gamma_j\tilde{A}_j = \delta_j\tilde{A} \qquad (115)$$

We shall call δ_j the *affinity factor* of step j. Instead of (110) we now have

$$\beta_a + \beta_c = \frac{x\delta_j}{\gamma_j} \qquad (116)$$

and instead of (111),

$$\alpha_{aj} = \frac{\beta_a\gamma_j}{x\delta_j} \qquad \alpha_{cj} = \frac{\beta_c\gamma_j}{x\delta_j} \qquad (117)$$

whereas (112–114) remain unchanged.

Whenever a Tafel behavior is observed, the interpretation by the MD method begins with the calculation of α_{aj} and α_{cj} from (113). We then obtain

$$\frac{x\delta_j}{\gamma_j} = \frac{\beta_a}{\alpha_{aj}} = \frac{\beta_c}{\alpha_{cj}} = \beta_a + \beta_c \qquad (118)$$

The number x is known from the way in which the overall electrode reaction has been written. We then obtain the ratio δ_j/γ_j. The stoichiometric number γ_j is a small integer whose value will often be obvious, in which case δ_j is obtained. With simple mechanisms δ_j should be close to and smaller than 1. The observed Tafel behavior implies that δ_j remains constant as i and η vary, and if δ_j is close to 1, that the fractions δ_k for the other steps of the mechanism, adding up to $1 - \delta_j$, do vary with i and η since, in the steady state and all these δ_k's being small, we have the expressions

$$i = i_k = i_{ok}\frac{(\beta_{ak} + \beta_{ck})F\eta}{RT} = \frac{i_{ok}\tilde{A}_k}{RT} = i_{ok}\frac{\delta_k}{\gamma_k}\frac{\tilde{A}}{RT} \qquad (119)$$

and we see that a constant δ_k would make i proportional to η in contradiction with the observed Tafel behavior.

Cases of mechanisms involving steps with affinity factors of the same order of magnitude as that of the rate-determining step can also occur. We shall see that cases in point are certain types of hydrogen overtension mechanisms.

The analysis of a Tafel case by means of the MD method includes the examination of the composition dependence of i_o. Formula (106) gives this dependence through $\tilde{\tilde{A}}_{je}$. This problem will be gone into in a later section.

6. Connection Between MD Kinetics and Absolute Rate Theory

We shall first consider the connection between the two forms of chemical kinetics for a heterogeneous, purely chemical reaction. It will be sufficient to examine only the forward rate. The MD formula for the forward velocity $\vec{v}_j$ of step j is

$$\frac{\vec{v}_j}{s} = \frac{1}{\lambda_j s}\exp\left(\frac{\vec{A}_j}{RT}\right) = \frac{\vec{v}_{je}}{s}\exp\left(\frac{\vec{A}_j - \vec{A}_{je}}{RT}\right) \qquad (120)$$

and the absolute rate formula is

$$\frac{\vec{v}_j}{s} = \vec{\varkappa}_j\frac{kT}{h}\vec{C}_{\pm} \qquad (121)$$

where $\vec{\varkappa}_j$ is a transmission coefficient of the order of unity for step j, kT/h is the Eyring fundamental frequency, and $\vec{C}_{\pm}$ is the concentration of the activated complex which is in near-equilibrium with the reactants of step j. Hence,

$$\vec{A}_j = \vec{\mu}_{\pm} = \vec{\mu}_{\pm}^0 + RT\ \ln(C_{\pm}f_{\pm}\varepsilon^2) \qquad (122)$$

where $f_{\pm}$ is the activity coefficient of the complex, and ε is the unit of length. Let us identify the two expressions of $\vec{v}_j/s$ for the case of equilibrium of the overall reaction. We regard $\vec{\varkappa}_j$ as varying with the instantaneous composition with a sensitivity of the same order as that of the activity coefficient of the complex. We then have

$$\frac{\vec{v}_{je}}{s} = \vec{\varkappa}_{je} \frac{kT}{h} \vec{C}_{\ast e} = \frac{1}{\lambda_j s} \exp\left(\frac{\vec{\mu}_{\ast}^0}{RT}\right) \vec{C}_{\ast e} f_{\ast e} \varepsilon^2 \qquad (123)$$

where the subscript e denotes equilibrium. Hence,

$$\frac{1}{\lambda_j s} = \vec{\varkappa}_{je} \frac{kT}{h} \exp\left(\frac{-\vec{\mu}_{\ast}^0}{RT}\right) \frac{1}{f_{\ast e} \varepsilon^2} \qquad (124)$$

and we see that our reasoning implies that $\vec{\varkappa}_j/\vec{f}_{\ast} = \vec{\varkappa}_{je}/\vec{f}_{\ast e}$.

Let us now consider the electrochemical case. We identify the MD and absolute-rate expressions at the electrode tension corresponding to a local value of zero for the electric potential at the site occupied by the activated complex, this state being indicated by the subscript 0. This means a state at which this electric potential is the same as that of the bulk solution. We then have, for a purely chemical reaction,

$$\frac{\vec{v}_{j0}}{s} = \vec{\varkappa}_{j0} \frac{kT}{h} \vec{C}_{\ast 0} \qquad (125)$$

Since the electric potential $\vec{\varphi}_{\ast}$ is zero,

$$\tilde{\vec{A}}_{j0} = \vec{\mu}_{\ast 0} = \vec{\mu}_{\ast}^0 + RT \ln(C_{\ast 0} \vec{f}_{\ast 0} \varepsilon^2) \qquad (126)$$

and the MD expression for $\vec{v}_{j0}/s$ is

$$\frac{\vec{v}_{j0}}{s} = \frac{1}{\lambda_j s} \exp\left(\frac{\vec{\mu}_{\ast}^0}{RT}\right) \vec{C}_{\ast 0} \vec{f}_{\ast 0} \varepsilon^2 \qquad (127)$$

Identifying (125) and (127) we obtain

$$\frac{1}{\lambda_j s} = \varkappa_{j0} \frac{kT}{h} \exp\left(\frac{-\vec{\mu}_{\ast}^0}{RT}\right) \frac{1}{\vec{f}_{\ast 0} \varepsilon^2} \qquad (128)$$

a formula providing fundamental information concerning the MD quantity λ_j. We now have, from (104) and (128),

$$\vec{i} = \frac{zF}{\gamma_j} \vec{\varkappa}_{j0} \frac{kT}{h} \exp\left(-\frac{\vec{\mu}_{\ast}^0}{RT}\right) \frac{1}{f_{\ast 0} \varepsilon^2} \exp\left(\frac{\vec{A}_j}{RT}\right) \qquad (129)$$

or*

$$\vec{i} = \frac{zF}{\gamma_j} \varkappa_{j0} \frac{kT}{h} \frac{f_{\ast}}{f_{\ast 0}} \exp\left(\frac{z_{\ast}F\vec{\varphi}_{\ast}}{RT}\right) \vec{C}_{\ast} \tag{130}$$

with similar expressions for $\overleftarrow{i}$, all arrows being reversed, whereas the original absolute rate theory would give us

$$\vec{i} = \frac{zF}{\gamma_j} \vec{\varkappa}_j \frac{kT}{h} \vec{C}_{\ast} \tag{131}$$

with a transmission coefficient given by

$$\vec{\varkappa}_j = \vec{\varkappa}_{j0} \exp\left(\frac{z_{\ast}F\vec{\varphi}_{\ast}}{RT}\right) \frac{\vec{f}_{\ast}}{\vec{f}_{\ast 0}} \tag{132}$$

which is now seen to be a function of $\vec{\varphi}_{\ast}$. Formula (129) can easily be transformed to bring out explicitly the standard chemical affinity of activation for the forward process $\vec{A}_{\ast}^0$:

$$\vec{i} = \frac{zF}{\gamma_j} \vec{\varkappa}_{j0} \frac{kT}{h} \exp\left(\frac{\vec{A}_{\ast}^0}{RT}\right) \frac{1}{f_{\ast 0}\varepsilon^2} \exp\left(\frac{\sum_i |v_i|F\varphi_i}{RT}\right) \Pi a_{i}^{|v_i|} \tag{133}$$

the summation Σ_i extending to all reactants. We shall come back to this expression in a later section.

7. Detailed MD Analysis of a Simple Electrode Process

In this section we shall treat in detail the reaction $X^{x+} + xe^- \rightarrow X$ already considered in section II.2 (see particularly formula (83)). We first assume that the rate-determining step is

$$X^{x+}(b') + xe^-(a') \rightarrow X(a'') \tag{134}$$

and that the electrochemical affinities of the various approach steps are all equal to zero. We then have

$$\tilde{\mu}_{X^{x+}}^{b'} = \tilde{\mu}_{X^{x+}}^c \qquad \tilde{\mu}_{e^-}^{a'} = \tilde{\mu}_{e^-}^a \qquad \mu_X^{a''} = \mu_X^a \tag{135}$$

*The quantity $[\exp(z_{\ast}F\vec{\varphi}_{\ast}/RT)]\vec{C}_{\ast}\vec{f}_{\ast}$ is of the type of an "electrochemical activity." Piontelli uses, for similar purposes, "electrochemical fugacities."

Hence, making i negative when $\eta < 0$,

$$i = -i_0 \left\{ \exp\left[\frac{\tilde{\mu}^c_{X^{x+}} + x\tilde{\mu}^a_{e^-} - (\tilde{\mu}^c_{X^{x+}e} + x\tilde{\mu}^a_{e^-e})}{RT} \right] - \exp\left(\frac{\mu^a_X - \mu^a_{Xe}}{RT} \right) \right\}$$

(136)

However,

$$\mu^a_X = \mu^a_{Xe} \qquad \tilde{\mu}^c_{X^{x+}} = \tilde{\mu}^c_{X^{x+}e}$$

(137)

and

$$\tilde{\mu}^a_{e^-} - \tilde{\mu}^a_{e^-e} = -F(\varphi^a - \varphi^a_e) = -F\eta$$

(138)

It follows that

$$i = -i_0 \left[\exp\left(\frac{-xF\eta}{RT} \right) - 1 \right]$$

(139)

This expression can be regarded as being of the Tafel type with

$$\beta_c = x, \beta_a = 0, \alpha_{cj} = 1, \alpha_{aj} = 0, \gamma_j = 1, |z| = x, \text{and } \delta_j = 1$$

Let us now assume that the approach step $X^{x+}(c) \to X^{x+}(b')$ has a small electrochemical affinity $\tilde{\mu}^c_{X^{x+}} - \tilde{\mu}^{b'}_{X^{x+}}$. This being step 1 in the cathodic mechanism we have (see section II.1)

$$v_1 = v = l_1(\tilde{\mu}^c_{X^{x+}} - \tilde{\mu}^{b'}_{X^{x+}})$$

(140)

and

$$\tilde{\mu}^c_{X^{x+}} - \tilde{\mu}^{b'}_{X^{x+}} = \frac{-i}{xFl_1}$$

(141)

Introducing this into (136) we have

$$i = -i_0 \left\{ \exp\left[-\frac{(i/xFl_1 + xF\eta)}{RT} \right] - 1 \right\}$$

(142)

or since the ratio i/xFl_1 is small,

$$i = -i_0 \left[\left(1 - \frac{i}{xFl_1 RT} \right) \exp\left(\frac{-xF\eta}{RT} \right) - 1 \right]$$

(143)

Hence, solving for i,

$$i = \frac{-i_0[\exp(-xF\eta/RT) - 1]}{1 - i_0 \exp(-xF\eta/RT)/xFl_1 RT}$$

(144)

If b' were to coincide with b, $\varphi^{b'}$ being then $\varphi^b = \zeta$ (electrokinetic potential) and the $\mu^0_{X^{x+}}$'s in b and c being equal to each other, formula (141) would give

$$i = -xFl_1 \left[RT \ \ln\left(\frac{a^c_{X^{x+}}}{a^b_{X^{x+}}}\right) - xF\varphi^b \right] \tag{145}$$

If b' is somewhat inside the transfer layer, the $\mu^0_{X^{x+}}$'s in b' and c are different from each other and we would have

$$i = -xFl_1 \left[\mu^{0c}_{X^{x+}} - \mu^{0b'}_{X^{x+}} + RT \ \ln\left(\frac{a^c_{X^{x+}}}{a^{b'}_{X^{x+}}}\right) - xF\varphi^{b'} \right] \tag{146}$$

This difference of $\mu^0_{X^{x+}}$'s can be regarded as a standard affinity of adsorption for ion X^{x+}.

We could similarly take into account a small difference $\tilde{\mu}^a_{e^-} - \tilde{\mu}^{a'}_{e^-}$ and also, in the anodic term, a small difference $\mu^{a''}_X - \mu^a_X$.

Let us note that i_0 for this case is given by

$$i_0 = \frac{xF}{\lambda_j s} \ \exp\left(\frac{\mu^a_X}{RT}\right) \tag{147}$$

and is thus independent of the composition of the solution.

It is of course possible that other steps besides that of discharge could involve appreciable fractions of the total electrochemical affinity and greater fractions than those of the approach steps which we have just considered. Dehydration of a hydrated ion, gradual penetration of X^{x+} through the transfer layer, entry of the deposited X into the metal lattice, etc. are such possibilities. These situations may still correspond to Tafel formulas, but with β_a and β_c different from 0 and x, respectively.

As an illustration of this point suppose we have obtained through anodic and cathodic polarization experiments $\beta_a = 0.36$ and $\beta_c = 0.54$, the charge number x being 1. We have $\alpha_{aj} = 0.4$ and $\alpha_{cj} = 0.6$. With $\gamma_j = 1$ we find $\delta_j = 0.9$. We would then conclude that

$$\tilde{A}_{ja} - \tilde{A}_{je} = 0.4 \ \tilde{A}_j = 0.4 \times 0.9 \ \tilde{A} = 0.36 \ \tilde{A} = 0.36 \ F\eta$$

$$\tilde{A}_{jc} - \tilde{A}_{je} = -0.6 \ \tilde{A}_j = -0.6 \times 0.9 \ \tilde{A} = -0.54 \ \tilde{A} = -0.54 \ F\eta$$

8. Case of Hydrogen Overtension

We shall assume that the cathodic reaction dominates. The mechanism will be written in that direction, but we shall write the expression for i in such a manner as to make $i < 0$.

The overall electrode process is

$$2 H^+(c) + 2e^-(a) \rightarrow H_2(a) \tag{148}$$

and its electrochemical affinity is $\tilde{A} = -2F\eta$, making $\tilde{A} > 0$ when $\eta < 0$.

We adopt the following mechanism:

$$
\begin{array}{ll}
H^+(c) \rightarrow H^+(b') & \text{I} \\[4pt]
e^-(a) \rightarrow e^-(a') & \text{II} \\[4pt]
H^+(b') + e^-(a') \rightarrow H(a'') & \text{III} \\[4pt]
2 H(a'') \rightarrow H_2(a''') & \text{IV} \\[4pt]
H_2(a''') \rightarrow H_2(a) & \text{V}
\end{array} \tag{149}
$$

Let us consider III as a rate-determining step (so-called "Volmer" mechanism). We have

$$i = - \frac{1}{\lambda_3 s} \left[\exp\left(\frac{\tilde{\mu}_{H^+}^{b'} + \tilde{\mu}_{e^-}^{a'}}{RT} \right) - \exp\left(\frac{\mu_H^{a''}}{RT} \right) \right] \tag{150}$$

If the electrochemical affinities of steps other than III can be taken as equal to zero, we have

$$i = -i_0 \left\{ \exp\left[\frac{\tilde{\mu}_{H^+}^c + \tilde{\mu}_{e^-}^a - (\tilde{\mu}_{H^+ + e}^c + \tilde{\mu}_{e^- - e}^a)}{RT} \right] - 1 \right\} \tag{151}$$

We neglect the mass transfer electrochemical affinity $\tilde{\mu}_{H^+}^d - \tilde{\mu}_{H^+}^c$. We thus have, as in (139),

$$i = -i_0 \left[\exp\left(\frac{-F\eta}{RT} \right) - 1 \right] \tag{152}$$

with

$$i_0 = \frac{F}{\lambda_3 s} \exp\left(\frac{\mu_{H_2}^a}{2 RT} \right) \tag{153}$$

independent of the composition of the solution. This is a Tafel case with $\beta_c = 1$, $\beta_a = 0$, $\alpha_{c3} = 1$, $\alpha_{a3} = 0$, $x = 2$, $\gamma_3 = 2$, and $\delta_3 = 1$.

Let us now assume that δ_3 is smaller than 1, with $(1 - \delta_3)\tilde{A}$ being distributed between steps I, II, IV, and V. We then have

$$\tilde{\mu}_{H^+}^{b'} - \tilde{\mu}_{H^+e}^{c} = \delta_1 F\eta$$

$$\tilde{\mu}_{e^-}^{a'} - \tilde{\mu}_{e^-e}^{a} = \tilde{\mu}_{e^-}^{a'} - \tilde{\mu}_{e^-}^{a} + \tilde{\mu}_{e^-}^{a} - \tilde{\mu}_{e^-e}^{a} = (\delta_2 - 1)F\eta \qquad (154)$$

$$\mu_H^{a''} - \tfrac{1}{2}\mu_{H_2}^{a} = -(\delta_4 + \delta_5)F\eta$$

and, instead of (151), we now have

$$i = -i_0\left\{\exp\left[\frac{(\delta_1 + \delta_2 - 1)F\eta}{RT}\right] - \exp\left[\frac{-(\delta_4 + \delta_5)F\eta}{RT}\right]\right\} \qquad (155)$$

If the δ's are all constant as η and i vary, this is an equation of the Tafel type with

$$\beta_c = 1 - (\delta_1 + \delta_2) \qquad \beta_a = -(\delta_4 + \delta_5) \qquad \beta_a + \beta_c = \delta_3$$
$$\alpha_{c3} = \frac{1 - (\delta_1 + \delta_2)}{\delta_3} \qquad \alpha_{a3} = -\frac{\delta_4 + \delta_5}{\delta_3} \qquad (156)$$

The appearance of negative values of the δ's or α's is a warning that the postulated mechanism needs revision.

Let us, e.g., consider a hypothetical case for which $\beta_c = \beta_a = \tfrac{1}{2}$, or $\delta_1 + \delta_2 = \tfrac{1}{2}$ and $\delta_4 + \delta_5 = -\tfrac{1}{2}$. This gives $\delta_3 = 1$, but some of the electrochemical affinities of the other steps are now different from zero. Let us assume that we have $\tilde{A}_1 = \tilde{A}_5 = 0$. We then have $\delta_2 = \tfrac{1}{2}$, $\delta_4 = -\tfrac{1}{2}$ and $\tilde{A}_2 = \tilde{A}/4$, $\tilde{A}_4 = -\tilde{A}/2$. We thus see that $\tilde{A}_3 + \tilde{A}_4 = 0$, which implies that at the reaction sites we have the electrochemical equilibrium

$$H^+(b') + H(a'') + e^-(a') \rightleftharpoons H_2(a''') \text{ or } \rightleftharpoons H_2(a) \qquad (157)$$

We have discussed this situation in detail in our earlier papers. The mechanism (149) should now be replaced by

$$H^+(c) \rightleftharpoons H^+(b') \qquad \tilde{A}_1 = 0$$

$$e^-(a) \rightarrow e^-(a') \qquad \tilde{A}_2 = \tilde{A}/4$$

$$H^+(b') + e^-(a') \rightarrow H(a'') \qquad \tilde{A}_3 = \tilde{A}/2 \qquad (158)$$

$$H^+(b') + H(a'') + e^-(a') \rightleftharpoons H_2(a''') \qquad \tilde{A}_4' = 0$$

$$H_2(a''') \rightleftharpoons H_2(a) \qquad \tilde{A}_5 = 0$$

We now have $\gamma_3 = 1$ and $2\tilde{A}_2 + \tilde{A}_3 = \tilde{A}$. If we now assume that in mechanism (158) applied to a particular metal-solution system step IV is rate-determining, we would be dealing with the so-called "Heyrovsky–Horiuti" mechanism. If we go back to mechanism (149) and consider step IV $[2\,H(a'') \rightarrow H_2(a''')]$ as rate-determining, we are dealing with the so-called "Tafel" mechanism. We easily see that a mechanism of the type (158), but with all $\tilde{A}$'s except $\tilde{A}'_4$ equal to zero, and a mechanism of the type (149), but with all $\tilde{A}$'s except $\tilde{A}_4$ equal to zero, give the same expression for the current:

$$i = -i_0\left[\exp\left(\frac{2\,\mu_H^{a''} - \mu_{H_2}^a}{RT}\right) - 1\right] = -i_0\left[\exp\left(\frac{-2F\eta}{RT}\right) - 1\right] \quad (159)$$

It seems likely that $\tilde{A}_2$ would depart from zero more in a system following the Heyrovsky–Horiuti mechanism than in a system following the Tafel mechanism since e^- is a reactant of the rate-determining step in the former mechanism and not in the latter. The experimental situation shows that, in the Volmer mechanism, β_a and β_c tend to differ from 0 and 1 and tend both to approach values around $\frac{1}{2}$, a situation for which the MD analysis strongly suggests that $\tilde{A}_2$, corresponding to a work of extraction of the electron, is then appreciably different from zero. The electron is a reactant in the rate-determining step both for the Volmer and the Heyrovsky–Horiuti mechanism. We could thus expect smaller variations from the factor 2 in the exponent of (159) in the Tafel case than in the Heyrovsky–Horiuti case. If we had $\tilde{A}_2 = \tilde{A}/4$ as in mechanism (158), this factor 2 would be reduced to 1. Changes away from 1 in the second term between brackets in (159) would depend on desorption affinities for H_2 and could be expected to be of the same order of magnitude in the two mechanisms.

The foregoing discussion of hydrogen overtension should be regarded more as an exercise in the use of the MD theory rather than as a strict and complete treatment for which space is not available here. We shall, however, come back to the problem in section (II.12) where we discuss the so-called Frumkin correction.

9. Composition Dependence of the Exchange Current i_0

From formula (129) we obtain for the exchange current (remembering that $|z| = x$),

$$i_0 = \frac{xF}{\gamma_j} \vec{\varkappa}_{j0} \frac{kT}{h} \exp\left(-\frac{\vec{\mu}_{\ne}^0}{RT}\right) \frac{1}{f_{\ne 0}\varepsilon^2} \exp\left(\frac{\tilde{\tilde{A}}_{je}}{RT}\right) \tag{160}$$

and i_0 is also equal to a similar expression in which all arrows are reversed. This implies the condition of identity:

$$\frac{\vec{\varkappa}_{j0}}{\vec{f}_{\ne 0}} \exp\left(\frac{-\vec{\mu}_{\ne}^0}{RT}\right) = \frac{\tilde{\varkappa}_{j0}}{\tilde{f}_{\ne 0}} \exp\left(\frac{-\tilde{\mu}_{\ne}^0}{RT}\right) \tag{161}$$

For the sake of simplicity and for easier comparison with current practice in electrode kinetics we shall take the simple redox reaction

$$\text{Red} \rightarrow \text{Ox} + xe^- \tag{162}$$

where Red and Ox are both in solution. We assume that the rate-determining step is

$$\text{Red}(a'') \rightarrow \text{Ox}(b') + xe^-(a') \tag{163}$$

where, for consistency with earlier portions of this treatment, the locations a', a'', and b' are similar to those already considered. We thus have, since φ^c is still taken as zero,

$$\tilde{\tilde{A}}_{je} = \tilde{\tilde{A}}_e = \mu_{\text{Red}}^{0c} + RT \ln(a_{\text{Red}}^c \varepsilon^3) \tag{164}$$

and hence, from (160),

$$i_0 = xF \frac{\vec{\varkappa}_{j0}\varepsilon}{\vec{f}_{\ne 0}} \frac{kT}{h} \exp\left(\frac{\mu_{\text{Red}}^{0c} - \vec{\mu}_{\ne}^0}{RT}\right) a_{\text{Red}}^c \tag{165}$$

where

$$\mu_{\text{Red}}^0 - \vec{\mu}_{\ne}^0 = \vec{A}_{\ne}^0 \tag{166}$$

is the standard chemical affinity of formation of the forward complex from the reduced species Red in c.

Similarly, in terms of the quantities pertaining to the right-hand side of (162),

$$\tilde{\tilde{A}}_{je} = \tilde{\tilde{A}}_e = \mu_{\text{Ox}}^{0c} + RT \ln(a_{\text{Ox}}^c \varepsilon^3) + x\mu_{e^-}^a - xF\varphi_e^a \tag{167}$$

and hence

$$i_0 = xF \frac{\tilde{\chi}_{j0} \varepsilon}{\tilde{f}_{\mp 0}} \frac{kT}{h} \exp\left(\frac{\mu_{Ox}^{0c} + x\mu_{e^-}^a - \bar{\mu}_{\pm}^0}{RT}\right) \exp\left(-\frac{xF\varphi_e^a}{RT}\right) a_{Ox}^c \qquad (168)$$

where

$$\mu_{Ox}^{0c} + x\mu_{e^-}^a - \bar{\mu}_{\mp}^0 = \bar{A}_{\pm}^0 \qquad (169)$$

is the standard chemical affinity of formation of the reverse complex from the oxidized species Ox in c and from the electron in a.

From (161), (165), and (168) we should have

$$\mu_{Red}^{0c} + RT \ln a_{Red}^c = \mu_{Ox}^{0c} + RT \ln a_{Ox}^c + x\mu_{e^-}^a - xF\varphi_e^a \qquad (170)$$

which is of course the necessarily true Nernst formula for electrochemical equilibrium:

$$\varphi_e^a = \frac{\mu_{Ox}^0 + x\mu_{e^-}^a - \mu_{Red}^0}{xF} + \frac{RT}{xF} \ln\left(\frac{a_{Ox}^c}{a_{Red}^c}\right) \qquad (171)$$

We thus see that our MD method is entirely self-consistent. We shall examine elsewhere the conditions under which our expressions (165) or (168) for i_0 may become compatible with expressions of the form $i_0 = kC_{Red}^{1-\alpha}C_{Ox}^{\alpha}$ which appear in most of the current presentations of electrode kinetics.

10. General Examination of the Components i_a and i_c of the Net Current Density i

We have, in general, whether a Tafel behavior is observed or not,

$$\frac{i_a}{i_0} = \exp\left(\frac{\tilde{A}_{ja} - \tilde{A}_{je}}{RT}\right) \qquad \frac{|i_c|}{i_0} = \exp\left(\frac{\tilde{A}_{jc} - \tilde{A}_{je}}{RT}\right) \qquad (172)$$

and we have examined in section II.10 the composition dependence of i_0.

Let us still consider the redox reaction (162) and its rate-determining step (163). We have

$$\tilde{A}_{Ja} - \tilde{A}_{je} = \mu_{Red}^{0a''} + RT \ln(a_{Red}^{a''}\varepsilon^2) + z_{Red}F\varphi^{a''}$$
$$- [\mu_{Red}^{0c} + RT \ln(a_{Red}^c\varepsilon^3)] \qquad (173)$$

and

$$
\tilde{A}_{jc} - \tilde{A}_{je} = \mu_{Ox}^{0b'} + RT \ln(a_{Ox}^{b'}\varepsilon^2) + z_{Ox}F\varphi^{b'} + x\tilde{\mu}_{e^-}^{a'} \\
- [\mu_{Ox}^{0c} + RT \ln(a_{Ox}^{c}\varepsilon^3) + x\tilde{\mu}_{e^- e}^{a}]
\tag{174}
$$

where the terms pertaining to the electron can be regarded as involving an electrochemical work of extraction $\tilde{w}_{e^-}$ such that

$$
\tilde{\mu}_{e^-}^{a'} - \tilde{\mu}_{e^- e}^{a} = \tilde{\mu}_{e^-}^{a'} - \tilde{\mu}_{e^-}^{a} + \tilde{\mu}_{e^-}^{a} - \tilde{\mu}_{e^- e}^{a} = \tilde{w}_{e^-} - F\eta
\tag{175}
$$

We may then rewrite (172) on the basis of (172–175)

$$
\frac{i_a}{i_0} = \exp\left(\frac{\mu_{Red}^{0a''} - \mu_{Red}^{0c}}{RT}\right) \frac{a_{Red}^{a''}}{a_{Red}^{c}\varepsilon} \exp\left(\frac{z_{Red}F\varphi^{a''}}{RT}\right)
\tag{176}
$$

and

$$
\frac{|i_c|}{i_0} = \exp\left(\frac{\mu_{Ox}^{0b'} - \mu_{Ox}^{0c}}{RT}\right) \frac{a_{Ox}^{b'}}{a_{Ox}^{c}\varepsilon} \exp\left(\frac{z_{Ox}F\varphi^{b'} + x\tilde{w}_{e^-} - xF\eta}{RT}\right)
\tag{177}
$$

Introducing the (165) expression of i_0 into (176) and the (168) one into (177) we have

$$
i_a = xF\frac{\vec{\varkappa}_{j0}}{\bar{f}_{\neq 0}} \frac{kT}{h} \exp\left(\frac{\mu_{Red}^{0a''} - \bar{\mu}_{\neq}^{0}}{RT}\right) a_{Red}^{a''} \exp\left(\frac{z_{Red}F\varphi^{a''}}{RT}\right)
\tag{178}
$$

and

$$
|i_c| = xF\frac{\vec{\varkappa}_{j0}}{\bar{f}_{\neq 0}} \frac{kT}{h} \exp\left(\frac{\mu_{Ox}^{0b'} - \bar{\mu}_{\neq}^{0}}{RT}\right) \\
\times a_{Ox}^{b'} \exp\left(\frac{x\mu_{e^-}^{a} + z_{Ox}\varphi^{b'} + x\tilde{w}_{e^-} - xF\eta}{RT}\right)
\tag{179}
$$

This latter formula can also be written as

$$
|i_c| = xF\frac{\vec{\varkappa}_{j0}}{\bar{f}_{\neq 0}} \frac{kT}{h} \exp\left(\frac{\mu_{Ox}^{0b'} + x\mu_{e^-}^{a'} - \bar{\mu}_{\neq}^{0}}{RT}\right) \\
\times a_{Ox}^{b'} \exp\left[\frac{z_{Ox}\varphi^{b'} - xF(\varphi^{a'} - \varphi_e^{a})}{RT}\right]
\tag{180}
$$

The developments in sections II.9 and II.10 show how the MD method handles cases other than those exhibiting Tafel behavior.

Let us note that the influence of adsorption, whether specific or nonspecific, and that of surface coverage can be examined on safe grounds by our MD method because such effects enter automatically in the expressions for the chemical potentials through both their standard portions and their activity terms. In addition, the electric terms in the electrochemical potentials are affected by all the circumstances of the structure of the interphase and of the mechanism of the process, but as shown in the foregoing section, these terms are never handled separately from the corresponding chemical terms. Assumptions of a mechanistic or molecular nature can be introduced with a reasonable degree of safety when the thermodynamic treatment has been carried out as far as has been done above for a typical simple case.

We shall now, for this same case, outline the treatment which is currently found in the literature.

11. Critique of the Customary Presentation of Electrode Kinetics

We shall again consider the reaction (162), and to simplify matters somewhat, we shall assume ideality in the solution and in the interphase. The problem is usually handled as follows: one first takes the particular case of a zero electric potential difference from metal to solution (from a to c in our notations) and the electrode process is then regarded as a purely chemical reaction of velocity

$$\frac{v}{s} = \vec{k}_0 \, C_{Red} - \overleftarrow{k}_0 \, C_{Ox} \tag{181}$$

If C_{Red} and C_{Ox} are taken as bulk concentrations in the solution, it is implied that equilibria are established for the approach steps from bulk to reacting sites. The dimensions of C_{Red} and C_{Ox} are (length)$^{-3}$ and it follows that $\vec{k}_0$ and $\overleftarrow{k}_0$ are in cm/sec if the cgs system of units is used.

If now an electric potential difference $\varphi^a - \varphi^c = \varphi^a$ (φ^c being taken as equal to zero) exists between metal and solution, the effect of the corresponding electric field on the reaction rate is assumed to be the transformation of the specific rates $\vec{k}_0$ and $\overleftarrow{k}_0$ to the following:

$$\vec{k} = \vec{k}_0 \, \exp\left(\frac{\alpha_a x F \varphi^a}{RT}\right) \qquad \overleftarrow{k} = \overleftarrow{k}_0 \, \exp\left[-\frac{(1 - \alpha_a) x F \varphi^a}{RT}\right] \tag{182}$$

where the parameters α_a and $\alpha_c = 1 - \alpha_a$ are introduced arbitrarily. We have here a stoichiometric number of 1. Let us note that no precise reference is made to the electron as reactant or product in the electrode process. It is clear that regardless of other arbitrary features, the reasoning implies a separation between chemical and electric terms of, at least, the electrochemical potential of the electron.

We thus write

$$\frac{v}{s} = \vec{k}_0 C_{\text{Red}} \, \exp\left(\frac{\alpha_a x F \varphi^a}{RT}\right) \ - \overleftarrow{k}_0 C_{\text{Ox}} \, \exp\left[-\frac{(1 - \alpha_a) x F \varphi^a}{RT}\right] \quad (183)$$

At electrochemical equilibrium we have $v = 0$ and

$$\vec{k}_0 C_{\text{Red}} \, \exp\left(\frac{\alpha_a x F \varphi_e^a}{RT}\right) \ = \overleftarrow{k}_0 C_{\text{Ox}} \, \exp\left[-\frac{(1 - \alpha_a) x F \varphi_e^a}{RT}\right] \quad (184)$$

However, according to the Nernst formula, we have

$$\exp\left[\frac{x F(\varphi_e^a - \varphi_e^0)}{RT}\right] = \frac{C_{\text{Ox}}}{C_{\text{Red}}} \quad (185)$$

and we thus have

$$\frac{\vec{k}_0}{\overleftarrow{k}_0} = \exp\left(-\frac{x F \varphi_e^0}{RT}\right) \quad (186)$$

Let us write separately

$$\vec{k}_0 = \vec{k} \, \exp\left(-\frac{\alpha_a x F \varphi_e^0}{RT}\right) \qquad \overleftarrow{k}_0 = \overleftarrow{k} \, \exp\left[\frac{(1 - \alpha_a) x F \varphi_e^0}{RT}\right] \quad (187)$$

and (183) becomes, in terms of the current density, with $\varphi = \varphi^a$:

$$i = \frac{x F v}{s} = x F k \left\{ C_{\text{Red}} \, \exp\left[\frac{\alpha_a x F(\varphi - \varphi_e^0)}{RT}\right] \right.$$

$$\left. - C_{\text{Ox}} \, \exp\left[-\frac{(1 - \alpha_a) x F(\varphi - \varphi_e^0)}{RT}\right] \right\} \quad (188)$$

a formula one finds in many electrochemical writings. Since

$$\varphi - \varphi_e^0 = \frac{RT}{xF} \, \ln\left(\frac{C_{\text{Ox}}}{C_{\text{Red}}}\right) \ + \eta \quad (189)$$

we also have

$$i = xFkC_{Red}^{1-\alpha_a}(C_{Ox}^{\alpha_a})\left\{\exp\left[\frac{\alpha_a xF\eta}{RT}\right] - \exp\left[-\frac{(1-\alpha_a)xF\eta}{RT}\right]\right\} \quad (190)$$

with

$$i_0 = xFkC_{Red}^{1-\alpha_a}(C_{Ox}^{\alpha_a}) \quad (191)$$

The reader is here referred to the final paragraph of section II.9.

12. MD Electrode Kinetics and the Frumkin Double Layer Correction

An interesting conflict arises between the implications of MD electrode kinetics and the derivation of what has become to be known as Frumkin's double-layer correction. Frumkin's derivation has remained essentially unchanged over the years between its first presentation in 1933[30] and, for instance, that of 1961 in "Advances in Electrochemistry and Electrochemical Engineering."[20] It can be summarized as follows, for the case of hydrogen overtension and assuming ideality:

When the step

$$H^+(b) + e^-(a') \rightarrow H(a'') \quad (192)$$

is rate-determining and when the anodic component of the current is negligible, the current density is taken as proportional to $C_{H^+}^b$, which is the factor taking into account the driving force due to H^+, and to a factor $\exp[-\alpha F(\varphi^a - \varphi^b)/RT]$ which, arbitrarily we submit, takes into account the driving force due to the electron. Frumkin[20] also introduces a free energy of desorption g_H of H from the metal surface which we may regard for our present purpose as constant and as incorporated into the specific rate k_F. Frumkin's formula is thus

$$i = -k_F C_{H^+}^b \exp\left[-\frac{\alpha F(\varphi^a - \varphi^b)}{RT}\right] \quad (193)$$

However, approach equilibrium is assumed for H^+ between c and b, and we thus have

$$\tilde{\mu}_{H^+}^b = \tilde{\mu}_{H^+}^c \qquad \text{or} \qquad C_{H^+}^b = C_{H^+}^c \exp\left(\frac{-F\varphi^b}{RT}\right) \quad (194)$$

since the $\mu_{H^+}^0$'s are the same in b and c. Combining with (193) we have

$$i = - k_F C_{H^+}^c \exp\left[\frac{-\alpha F \varphi^a - (1 - \alpha)F\varphi^b}{RT}\right] \tag{195}$$

The factor $\exp[-(1 - \alpha)F\varphi^b/RT]$ represents the Frumkin correction.

In MD kinetics, however, we have in place of (193)

$$i = - \frac{1}{\lambda s} \exp\left(\frac{\tilde{\mu}_{H^+}^{b'} + \tilde{\mu}_{e^-}^{a'}}{RT}\right) \tag{181}$$

and position b' may not coincide with b. If we retain the assumption of approach equilibrium for H^+ we have

$$\tilde{\mu}_{H^+}^{b'} = \mu_{H^+}^0 + RT \ln(C_{H^+}^c \varepsilon^3) \tag{197}$$

and we have

$$i = - k C_{H^+}^c \exp\left(\frac{\tilde{\mu}_{e^-}^{a'}}{RT}\right) \tag{198}$$

Let us introduce the electrochemical work $\tilde{w}_{e^-}$ of extraction of the electron from the bulk of the metal to the reacting site a':

$$\tilde{w}_{e^-} = \tilde{\mu}_{e^-}^{a'} - \tilde{\mu}_{e^-}^{a} \tag{199}$$

It follows that

$$\tilde{\mu}_{e^-}^{a'} = \tilde{w}_{e^-} + \mu_{e^-}^{a} - F\varphi^a \tag{200}$$

and from (198) we obtain

$$i = - k' C_{H^+}^c \exp\left(\frac{\tilde{w}_{e^-} - F\varphi^a}{RT}\right) \tag{201}$$

In order to compare formulas (195) and (201) let us examine the implied exchange currents:

$$i_{OF} = k_F C_{H^+}^c \exp\left(\frac{-\alpha F \varphi_e^a - (1 - \alpha)F\varphi_e^b}{RT}\right) \tag{202}$$

$$i_{OMD} = k' C_{H^+}^c \exp\left(\frac{-F\varphi_e^a}{RT}\right) \tag{203}$$

According to the Nernst formula we have

$$\varphi_e^a = \varphi^0 + \frac{RT}{F} \ln C_{H^+}^c \tag{204}$$

and the exchange currents become

$$i_{OF} = k_F'(C_{H^+}^c)^{1-\alpha} \exp\left[-\frac{(1-\alpha)F\varphi_e^b}{RT}\right] \tag{205}$$

$$i_{OMD} = k'' \tag{206}$$

We see that i_{OF}, like i_{OMD}, will reduce to a constant if

$$\varphi_e^b = \varphi_e^a + \text{constant} \tag{207}$$

a relation which is apparently well verified whenever the solution contains a single electrolyte.

From (195) and (205) we obtain

$$i = -i_{OF} \exp\left(\frac{-\alpha F\eta - (1-\alpha)F(\varphi^b - \varphi_e^b)}{RT}\right) \tag{208}$$

and from (201) and (206) we obtain

$$i = -i_{OMD} \exp\left(\frac{\tilde{w}_{e^-} - F\eta}{RT}\right) \tag{209}$$

Considering a particular composition (in order to be able to regard i_{OF} as constant) we see that these two expressions for i will describe the same experimental situation if we identify the exponents of (208) and (209). We then have

$$\tilde{w}_{e^-} = (1-\alpha)F[\eta - (\varphi^b - \varphi_e^b)] = (1-\alpha)F[\varphi^a - \varphi^b - (\varphi_e^a - \varphi_e^b)] \tag{210}$$

There does not seem to be any obvious reason to regard $\tilde{w}_{e^-}$ as being correctly given by this expression and this identification does not remove, at this stage of the reasoning, the conflict between the Frumkin and MD treatments. We shall give elsewhere a detailed study of this problem. Let us take the particular case $\varphi^b = \varphi_e^b$ which would be realized in presence of a sufficient amount of supporting electrolyte. We would then have

$$\tilde{w}_{e^-} = (1-\alpha)F\eta \tag{211}$$

If the approach process $e^-(a) \rightarrow e^-(a')$ were to have a zero electrochemical affinity, we would have $\tilde{w}_{e^-} = 0$ and according to (201),

$$i = -i_{0MD} \; \exp\left(\frac{-F\eta}{RT}\right) \tag{212}$$

and, when $\tilde{w}_{e^-}$ is different from zero, this quantity would then be the fraction $1 - \alpha$ of $F\eta$.

In our opinion no Frumkin correction should here be made when the approach process $H^+(c) \rightarrow H^+(b)$ is assumed to have an electrochemical affinity of zero. The contribution of $H^+(b)$ to the reaction velocity is its electrochemical activity $C_{H^+}^b \exp(F\varphi^b/RT)$ $= C_{H^+}^c$, not its concentration $C_{H^+}^b$.

Corrections depending on the features of the electrochemical double layer will be applicable to the factor $\exp(-\alpha F\eta/RT)$ (once this factor has been introduced *a priori*) when the approach process $H^+(c) \rightarrow H^+(b)$ is not at electrochemical equilibrium, when the process of extraction of the electron also involves a nonzero electrochemical affinity, when a nonequilibrium adsorption process precedes discharge, etc.

IV. CONCLUSIONS

Not only the advisability, but also the actual necessity of using thermodynamics as far as it will penetrate into the intricacies of electrochemical phenomena should be demonstrated by the fore-going discussion.

Besides those mentioned in the Introduction we have left out of our presentation a number of topics: some which can easily be derived from what has been given in our treatment, such as the examination of the temperature variation of i's and i_0's and of the related enthalpies of activation; some which, although they are of very great interest and likely to receive considerable clarification from the thermodynamics of irreversible processes, would involve for the moment too much speculation, such as the study of simultaneous overall electrode processes occurring near or far from their reversible electric tensions. In the case, e.g., of two electrode processes with neighboring *e.m.f.*'s, linear relations between partial currents and overtensions could be set up and tested experimentally from the point of view of the correctness of the Onsager symmetry relation between mutual influence coefficients. In fact the remarkable advances in electrochemical experimentation make possible precise investigations in the range of small departures from equilibrium, whereas the corresponding situation with purely chemical reactions is much less favorable. As a whole, electrochemistry is an ideal domain for the application and the testing of all the resources of thermodynamic reasoning.

REFERENCES

[1] Agar, *Thermogalvanic Cells* in *Advances in Electrochemistry and Electrochemical Engineering*, Ed. Delahay and Tobias, Vol. 3, pp. 31–121, Interscience Publishers Inc. and J. Wiley & Sons, New York, 1963.

[2] Van Rysselberghe, *Electrochemical Affinity—Studies in Electrochemical Thermodynamics and Kinetics*, Hermann, Paris, 1955; Lange and Göhr, *Thermodynamische Elektrochemie*, Hüthig Verlag, Heidelberg, 1962.

[3] Van Rysselberghe, *J. Chem. Phys.* **17** (1949) 1229.

[4] Van Rysselberghe, *J. Chem. Phys.* **29** (1958) 640.

[5] Marcelin, *Ann. phys.* **3** (1915) 158.

[6] De Donder, *L'Affinité*, Gauthier-Villars, Paris, 1927.

[7] Van Rysselberghe, *Atti Accad. Naz. Lincei, Rend. Classe Sci. Fis. Mat. Nat., Ser. VIII*, **31** (1961) 391.

[8] Van Rysselberghe, in *Atlas d'Equilibres Electrochimiques à 25°C* by M. Pourbaix *et al*, pp. 21–27, Gauthier-Villars, Paris, 1963.

[9] Van Rysselberghe, *Electrochim. Acta* **8** (1963) 583.

[10] Van Rysselberghe, *Electrochim. Acta* **8** (1963) 709.

[11] Van Rysselberghe, *Electrochim. Acta* **9** (1964) 1547.

[12] Gutmann, and Van Rysselberghe, *Electrochim. Acta* **10** (1965) 107.

[13] Van Rysselberghe, *Thermodynamics of Irreversible Processes*, Blaisdell Publishing Co. New York and Hermann, Paris, 1963.

[14] De Donder, and Van Rysselberghe, *L'Affinité*, Gauthier-Villars, Paris, 1936. Also *Thermodynamic Theory of Affinity*, Stanford University Press, Stanford, 1936.

[15] Prigogine, *Etude Thermodynamique des Phénomènes Irréversibles*, Desoer, Liège and Dunod, Paris, 1947.

[16] Prigogine, *Introduction to Thermodynamics of Irreversible Processes*, C. C. Thomas, Springfield, Illinois, 1955. 2nd ed., Interscience Publishers Inc. and J. Wiley & Sons, New York, 1961.

[17] Prigogine, and Defay, with Everett, *Chemical Thermodynamics*, Longmans Green, London, New York, and Toronto, 1954.

[18] Vetter, *Elektrochemische Kinetik*, Springer Verlag, Berlin, 1961.

[19] Parsons, *The Structure of the Electrical Double Layer and its Influence on the Rates of Electrode Reactions* in *Advances in Electrochemistry and Electrochemical Engineering*, Ed. Delahay and Tobias, Vol. 1, pp. 1–64, Interscience Publishers Inc. and J. Wiley & Sons, New York, 1961.

[20] Frumkin, *Hydrogen Overvoltage and Adsorption Phenomena*, Part I, in *Advances in Electrochemistry and Electrochemical Engineering*, Ed. Delahay and Tobias, Vol. 1, pp. 65–121, Interscience Publishers Inc. and J. Wiley & Sons, New York, 1961.

[21] Frumkin, *Hydrogen Overvoltage and Adsorption Phenomena*, Part II, in *Advances in Electrochemistry and Electrochemical Engineering*, Ed. Delahay and Tobias, Vol. 3, pp. 287–391, Interscience Publishers Inc. and J. Wiley & Sons, New York, 1963.

[22] Gierst, *Cinétique d'Approche et Réactions d'Electrodes Irréversibles*, Buteneers, Liège, 1958.

[23] Hurwitz, *Contribution à la Détermination des Paramètres Cinétiques Réels des Processus d'Electrode*, Thesis, Brussels, 1964.

[24] Piontelli, see: *Comptes Rendus de la 2ème Réunion du CITCE*, pp. 79–104, Tamburini, Milan, 1951; *Gazz. Chim. Ital.* **85** (1955) 665; *Chim. Ind. Milan* **35** (1953) 421.

[25] Van Rysselberghe, *Proceedings of 9th Meeting of CITCE*, pp. 176–219, Butterworths, London, 1959.

[26] Van Rysselberghe, CITCE Report on *Electrochemical Nomenclature and Definitions*: *Electrochim. Acta* **5** (1961) 28; **8** (1963) 543; **9** (1964) 1343; *J. Electroanal. Chem.* **2** (1961) 265; **6** (1963) 173; **7** (1964) 417. With Lange, *J. Electrochem. Soc.* **105** (1958) 420.

[27] Christiansen, and Pourbaix, *Comptes Rendus de la 17ème Conférence de l'Union Internationale de Chimie Pure et Appliquée (IUPAC)*, Stockholm, (1953), p. 82, Maison de la Chimie, Paris, 1954. Christiansen, *J. Amer. Chem. Soc.* **82** (1960) 5517.

[28] Van Rysselberghe, *Electrochim. Acta* **3** (1961) 257.

[29] Van Rysselberghe, *J. Chem. Educ.* **41** (1964) 486; see also *Thermodynamics of Irreversible Electrochemical Processes*, pp. 743–747, and *Remarks on Nomenclature*, pp. 859–861, in *The Encyclopedia of Electrochemistry*, Ed. Hampel, Reinhold Publishing Corp., New York, 1964.

[30] Frumkin, *Z. Physik. Chem.* **A164** (1933) 121.

2

The Mechanism of Electrochemical Oxidation of Organic Fuels

B. J. Piersma and E. Gileadi

I. INTRODUCTION

1. Scope

The recent interest in electrochemical energy conversion[1] and the discovery by Heath and Worsham[2] that saturated hydrocarbons, which are potentially economical fuels, could be completely oxidized spontaneously with the production of electrical energy, have led to renewed efforts in the study of the electrochemical oxidation of organic compounds. Since much of the work in electro-organic chemistry has not been conducted with a clear philosophy of mechanism determination, any discussion of electrode kinetics in this area is necessarily severely limited. This discussion will thus attempt to clarify the type of information which is required for mechanistic interpretation, the experimental methods by which this information is most profitably obtained, and the treatment of this information once it has been obtained. The large field of organic polarography, both the conventional reduction of organic compounds and oxidation at solid electrodes, will not in general be discussed here, since the primary aim in that area is not mechanism determination but electrochemical analysis and is properly discussed in terms of electroanalytical chemistry. Electrosynthesis will also not be examined since, at the present time, this field is more of an art than a science, although "future research in this field has a much brighter prospect, not only because electrode kinetic concepts are now becoming well known, but because of the availability of potentiostatic controlled devices."[3]

2. Brief Historical Review

The field of electro-organic chemistry has been little developed from an electrochemical point of view and even the simplest electro-organic reactions have not been very satisfactorily formulated, although this was early a field of much activity in electrochemistry. This state of affairs may be partially attributed to the complexity of the reactions. However, the quality of the work carried out, particularly in electrosynthetic investigations, leaves much to be desired. For example, as early as 1905, it could be stated that "A very considerable amount of work has been done upon the electrolytic oxidation of organic substances, but most of the work has been of rather a fragmentary style, and the substances to be dealt with have, to a large extent, been apparently picked out at haphazard. Therefore, in spite of the work done, very little is known of the actual conditions necessary for carrying out electrolytic oxidation, and in what way it differs from purely chemical oxidation."[4] A similar statement would have been justified until very recently, and to a large extent is justified even at the present time. Thus, in most reported electro-organic oxidations up to recent times, the percentage yields and experimental conditions have not been indicated or controlled and therefore many of the studied reactions should be reinvestigated under more rigorous electrochemical control.

The idea of using electricity as an oxidizing or reducing agent was formulated early in the nineteenth century by Reinhold and Erman, who electrolyzed aqueous solutions of alcohols.[5] The first detailed study of products obtained using different electrodes for alcohol oxidation was made by Ludersdorff in 1830.[6] Although Faraday had made earlier the observation that some hydrocarbons could be obtained upon electrolysis of acetate solutions,[7] it was Kolbe who made the first applications of electrolytic oxidation for electrosynthesis when he obtained ethane from the electrolysis of alkali acetates.[8] Kolbe's studies led to extensive investigations of the electrolysis of aromatic hydrocarbons and their derivatives. However, many researchers were discouraged by the fact that, "Practically in every electrochemical oxidation of an organic compound, part of the organic compound is oxidized to carbon dioxide and water, because electrolytic or atomic oxygen is the most effective oxidizing agent that we possess."[9] Most of the following investigations were concerned with finding conditions under which the organic compound was not completely oxidized and useful intermediate products could be obtained. (Note the opposite requirement for electrolytic oxidation in fuel cells where the highest efficiency is obtained from complete oxidation to carbon dioxide.) In 1859, Friedel electrolytically oxidized acetone and found a mixture of formic, acetic, and carbonic acids with evolution of carbon dioxide and oxygen at the anode in an acetone-sulfuric acid mixture.[10] Further studies on ketones were not reported until 1931, when a similar study was carried out resulting in the formation of methane, ethane, and unsaturated hydrocarbons, in addition to carbon dioxide and oxygen at platinum anodes.[11] The first anodic oxidation of benzene was reported in 1880, with the observation that the electrolytic oxidation of benzene in an ethanolic-sulfuric acid medium yielded unidentifiable substances.[12] A few years later Gotterman and Friedrichs reported that hydrocarbons were obtained from the anodic oxidation of benzene in alcoholic-sulfuric acid solution at platinum anodes.[13]

At about the same time, in his attempts to imitate biochemical reactions using alternating current electrolysis of phenol in weakly acid sulfate solutions at platinum electrodes, Drechsel reported many products including diphenyl derivatives formed by nuclear linkage.[14] Drechsel was primarily interested in producing urea by electrolysis of amino acids and comparing this with physiological chemical processes. Using several different electrodes, e.g., platinum, palladium, and copper, he found that a number of compounds, such as hydroquinone, pyrocatechol, succinic, oxalic, and formic acids, were obtained and that a part of the reactant "disappeared," probably being completely oxidized to carbon dioxide which he did not detect. This type of approach was continued, e.g., Nathansohn was interested in "capillary-electric processes in living cells" and potential differences in living organisms.[15] Fichter, who was very active in the electro-organic field, however, advised caution in a comparison of biochemical and electrochemical oxidation of organic compounds.[16]

In 1898, in a study of the cathodic reduction of nitrobenzene, Haber discovered that by varying the electrode potential he could obtain either azoxybenzene or hydrazobenzene. This led him to make the following significant conclusions: "The electric current up to this time has been regarded in organic electrochemistry as a means of reaction whose effects are determined through current density, current duration and occasionally through the material of the electrode. This view is incomplete, for oxidation and reduction processes depend mainly on the potential of the electrode at which they take place. The current density, current duration and electrode material are important only in so far as they determine the electrode potential and its changes in the process of electrochemistry."[17] This fact was almost completely overlooked by subsequent investigators of electro-organic reactions, until the invention by Hickling in 1942[18] of a device for automatically controlling the potential of a working electrode at a desired value, called a potentiostat.

In the early twentieth century, numerous investigations in electro-organic synthesis were made. However, all of these were incomplete and not primarily aimed at mechanism determination. One of the leading investigators of electro-organic reactions during this period was Fichter, who began a study of benzene and its homologs around 1910. These included several aromatic compounds such as phenol, toluene, xylene, and cymene. A review of his investigations on aromatic hydrocarbons was presented to the Electrochemical Society in 1924.[19] In his treatment of the Kolbe reaction, Fichter supported the aryl peroxide mechanism,[20] one of the three theories proposed for the mechanism of the Kolbe electrosynthesis which have received serious consideration (the other two mechanisms, the hydrogen peroxide theory and the discharged ion or free radical mechanism will be presented subsequently). As originally postulated by Schall,[21] this theory maintained that active oxygen liberated at the anode oxidized the carboxylic acid anion which then decomposed through an acyl peroxide intermediate. Fichter later proposed that the primary step was discharge from the carboxylic acid anions followed by combination forming diacyl peroxide intermediates which decomposed with formation of the characteristic Kolbe products.[22] According to this theory, the reaction in alkaline solution may be written:

$$2RCOO^- \rightarrow 2RCOO\cdot + 2e \qquad (1)$$

$$2RCOO\cdot \rightarrow R-\overset{\overset{\displaystyle O}{\displaystyle \|}}{C}-O-O-\overset{\overset{\displaystyle O}{\displaystyle \|}}{C}-R \qquad (2)$$

$$R\text{—}\overset{\overset{\displaystyle O}{\|}}{C}\text{—}O\text{—}O\text{—}\overset{\overset{\displaystyle O}{\|}}{C}\text{—}R \rightarrow 2R\cdot + 2CO_2 \tag{3}$$

$$R\cdot + R\cdot \rightarrow R\text{—}R \tag{4}$$

In the 1920's, E. Müller and his co-workers made a series of studies on the anodic oxidation of methanol, formaldehyde, and formic acid which represent the first extensive mechanistic investigation of these compounds,[23-28] although the principles of electrode kinetics had not yet been formulated. Müller did not establish mechanisms for these reactions; however, many of his observations have been later confirmed and his studies were among the first with a comparison of polarization curves on several noble metals including platinum, palladium, rhodium, iridium, osmium, rubidium, gold, and silver (cf. Figure 1). As was usual at that time, Müller discussed his results in terms of polarization, rather than in terms of current or reaction rate.

Following this were the investigations of Glasstone and Hickling, centered primarily on the Kolbe and Hofer-Moest reactions

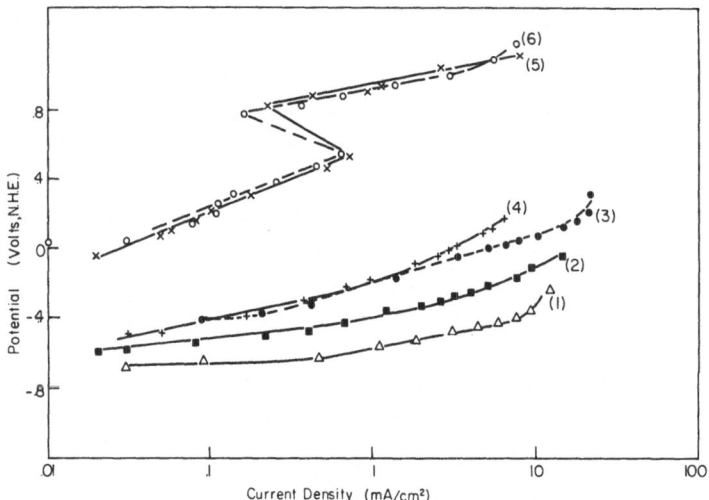

Figure 1. Polarization curves for the anodic oxidation of formate showing the influence of the catalyst: (1) palladized Pd, 20°C, (2) platinized Pt, 20°C, (3) Pd, 75°C, (4) rhodenized Pt, 20°C, (5) Pt, 75°C, (6) Ir, 75°C [from E. Müller, *Z. Elektrochem.* **29** (1923) 264].

in aqueous solutions, resulting in their hydrogen peroxide theory.[29-31] Their mechanism assumed the combination of discharged hydroxyl radicals to form hydrogen peroxide which then initiated the decomposition of the carboxylate ions. The reaction may be formulated according to the hydrogen peroxide theory as

$$2OH^- \rightarrow 2OH\cdot + 2e \qquad (5)$$

$$2OH\cdot \rightarrow H_2O_2 \qquad (6)$$

$$H_2O_2 + 2RCOO^- \rightarrow 2RCOO\cdot + 2OH^- \qquad (7)$$

$$2RCOO\cdot \rightarrow R{-}R + 2CO_2 \qquad (8)$$

The Hofer-Moest reaction, which is the formation of an alcohol rather than the hydrocarbon, was considered to occur whenever the conditions were not favorable for the Kolbe reaction:

$$H_2O_2 + RCOOH \rightarrow R\overset{\overset{\displaystyle O}{\displaystyle \|}}{C}{-}O{-}OH + H_2O \qquad (9)$$

$$R\overset{\overset{\displaystyle O}{\displaystyle \|}}{C}{-}O{-}OH \rightarrow ROH + CO_2 \qquad (10)$$

Other investigators have not seriously considered this mechanism primarily because hydrogen peroxide has not been detected in the electrolytic solution. Their work, however, might be considered a forerunner to the present water discharge mechanism for the electrochemical oxidation of unsaturated hydrocarbons as suggested by Wroblowa, Piersma and Bocknis.[32,33] Glasstone and Hickling also contributed one of several books on the subject of organic electrochemistry[34] (others are by Fichter,[35] Allen,[6] and Brockman[36]).

The Kolbe-type reaction has continued to receive much attention with particular emphasis on formic acid oxidation. However, the mechanism remains open to question.[37-39] The mechanism given primary consideration at present is that originally suggested by Crum-Brown and Walker,[40] who proposed direct electrochemical oxidation of the carboxylate ion with subsequent decomposition of the radicals which combine giving the hydrocarbon. This theory, known as the discharged ion and free radical mechanism, can be given in the general form

$$2RCOO^- \rightarrow 2RCOO\cdot + 2e \qquad (1)$$

$$2RCOO\cdot \rightarrow 2R\cdot + 2CO_2 \qquad (11)$$

$$R\cdot + R\cdot \rightarrow R—R \qquad (4)$$

Although by 1900 the electrochemical oxidation of aromatic hydrocarbons was extensively examined, the aliphatic hydrocarbons received little attention; probably either because of the anticipated difficulty of oxidizing saturated hydrocarbons or their complete oxidation to carbon dioxide which, until recent times, were not reactions of interest. The first report of the use of ethylene in an electrochemical reaction was in 1916 in a study of the chlorination of ethylene.[41] Using platinum electrodes in 0.1 N HCl, saturated with ethylene and chlorine in compartments separated by a stopcock, these authors obtained chlorination of the double bond. They assumed that ethylene lost two electrons to the electrodes and then combined with two chloride ions. The next recorded study of the electrolytic oxidation of ethylene was in 1946, when no oxidation was found at lead anodes. However, $(CH_2OH)_2$ was formed at platinum under special conditions (7° and a current density of 70 A/cm^2).[42] In 30% acetone–aqueous KOH solutions, both ethylene and acetone were oxidized to $(CH_2OH)_2$, HCHO, CH_3CHO, and other products. No attempt was made to discuss mechanism. Other oxidations of ethylene have been claimed in several patents. The product at platinum[43] and carbon[44] electrodes in strong alkaline solutions was stated to be ethylene glycol, while oxidation at lead electrodes in saturated sulfuric acid was said to yield carboxylic acids.[45,46] The oxidation of paraffinic hydrocarbons at activated lead anodes in sulfuric acid at room temperatures to form carboxylic acids has also been reported.[47–51] More recently, several studies of the electrochemical oxidation of hydrocarbons have appeared in the literature,[52–68] although in most cases either they have not been primarily directed toward determining the usual parameters of electrode kinetics useful for diagnosing mechanisms or they have not presented sufficient evidence to allow valid mechanistic conclusions. The few exceptions[2,32,33,69,70] will be discussed in detail in a later section.

Many other organic compounds have been examined at anodes. However, until about 1950 few mechanism studies were reported. In 1930, the anodic oxidation of acetic acid in sulfuric acid at platinum electrodes was reported to proceed via the following process:[71]

$$CH_3COOH \xrightarrow{\ 0\ } H_2COH—COOH \xrightarrow{\ 0\ } HC(OH)_2—COOH \quad (12)$$

$$HC(OH)_2—COOH \xrightarrow{-2H} (COOH)_2 \xrightarrow{-2H} 2CO_2 \quad (13)$$

Side reactions reportedly formed formaldehyde and formic acid. Glucose and other sugars have been oxidized by AC current electrolysis at platinum and graphite electrodes[72] and were reported to be degraded in a step-wise fashion quantitatively to carbon dioxide.[73] Even cellulose has been found to be oxidized electrochemically with a reported oxidation of the hydroxyl groups in the pyranose rings and hydrolysis of the glucoside linkages.[74] In another study of cellulose at platinum anodes in $1 N H_2SO_4$, "considerable degradation" was observed. However, cellulose appeared to retain its physical strength and fibrous character.[75] The complete oxidation of cellulose and lower carbohydrates to carbon dioxide and a comparison of polarization curves has also recently been reported.[76]

Beginning with the paper of Pavela in 1954 on the low-potential oxidation of methanol at platinized platinum electrodes,[77] the investigations of electro-organic oxidations have been primarily directed toward mechanism studies with particular emphasis on formic acid, formaldehyde, and methanol as the simplest reactions which can be studied. (Actually the simplest electro-organic oxidation is that of CO to CO_2, which may be considered the organic analog of the hydrogen evolution reaction in electrochemistry.) Unfortunately, most of the investigations since that time have been made using non-steady-state (transient) methods such as chronopotentiometry and potential-scanning techniques (solid electrode polarography or cyclic voltammetry). In general only qualitative information has hitherto been obtained by these methods from which reliable mechanistic conclusions have either not been made or could not be made. One exception to this is in certain cases where reversible reactions are examined under conditions of diffusion control.

Reinmuth has examined chronopotentiometric potential–time curves and proposed diagnostic criteria for their interpretation.[78] His treatment applies to the very limited cases with conditions of semi-infinite linear diffusion to a plane electrode, where only one electrode process is possible and where both oxidized and reduced forms of the electroactive species are soluble in solution. This approach is further restricted in application, in many cases, to electrode processes whose rates are mass-transport controlled. Nicholson and Shain have examined in some detail the theory of stationary electrode polarography for single-scan and cyclic methods applied to reversible and irreversible systems.[79] However, since in kinetic studies it is preferable to avoid diffusion control which obscures the reaction kinetics, such methods are not well suited for the general study of the mechanism of electrochemical organic oxidation. The relatively few studies which have attempted to analyze the mechanisms of electrochemical organic oxidation reactions will be discussed in detail in a following section.

It may be concluded from a survey of the electro-organic literature that much of the work is filled with errors, partly due to misconceptions and partly due to incorrect observations, since most of the research, even at the present time, has been carried out in a very superficial manner.

II. EXPERIMENTAL METHODS FOR ELUCIDATION OF MECHANISMS OF ELECTRO-ORGANIC OXIDATION

1. General

Anodic processes have, in general, received less attention than cathodic processes because of characteristic experimental difficulties which are absent in most cathodic studies. Most metals undergo anodic dissolution. Mercury, for example, which has been used extensively in cathodic reduction studies, is not suitable for the study of anodic reactions. Investigations of most anodic processes have so far been limited to the noble metals and carbon, although the possibility of studying such reactions on oxide surfaces or on refractory intermetallic compounds appears promising. The study of anodic organic oxidations is further limited to a potential region between the open-circuit potential of the organic system which usually is 0.2–0.3 V positive to the reversible hydrogen potential, and the potential of oxide formation (approximately 1 V on the reversible hydrogen scale).*

The quantitative and theoretically significant approaches to the study of the electrode kinetic reactions have been established only within the last twenty years.[80-85] These methods have been applied in the study of only a few electrochemical systems, particularly the hydrogen and oxygen evolution reactions. Further, these approaches have been seldom applied for the elucidation of mechanisms of the anodic oxidation of organic compounds. The recent literature indicates a trend toward methods which give large amounts of data in relatively short times but which do not lend themselves easily to quantitative analysis. Thus, it seems desirable at this point to review the experimental methods which may be profitably used in the study of organic reactions at anodes.

In this discussion it will be useful to distinguish between the methods of electroanalytical studies[86-88] and those more specifically designed for mechanism studies. The latter involve: (a) Elucidation of the reaction mechanism; (b) identification of the rate-limiting step; (c) derivation of adsorption isotherms and behavior of the adsorbed reactants; and (d) the examination of adsorbed intermediates.

2. Overall Reaction

The overall reaction is characterized by product analyses and coulombic efficiency determinations. Carbon dioxide, which is the primary product in most organic reactions at noble-metal anodes, can be removed from acidic solutions in the anode compartment of the electrolytic cell by passing an inert gas through the cell, and then reacting quantitatively, e.g., with $Ba(OH)_2$ or "Ascarite." Carbon dioxide is not as easily determined in alkaline electrolyte, which must be analyzed, e.g., by volumetric methods or chromatography. Nonvolatile products from the oxidation are determined by analysis of the electrolyte, e.g., by gas chromatography, preferably with a flame ionization detector, or mass spectroscopy. Organic products can be extracted from the electrolyte

---

*The formation of oxide on the electrode usually prevents further significant oxidation of the organic species. However, at increased potentials the organic may be oxidized on the oxide along with oxygen evolution. A notable exception is the Kolbe decarboxylation reaction which occurs at potentials well above the oxygen reversible potential without oxygen evolution.

by standard procedures and analyzed by infrared or ultraviolet spectroscopy, etc. Volatile products can be analyzed directly on a gas chromatograph or condensed and analyzed by any of the usual methods. The coulombic efficiency or current efficiency of a reaction is established by comparison of the quantity of product obtained experimentally with the number of coulombs passed during the reaction. This is often done at constant current, eliminating the necessity of integration. However, since the coulombic efficiency may be a function of electrode potential,[89] it is sometimes preferable to work potentiostatically.

3. Determination of Equilibrium Adsorption

In most electrochemical reactions, except very fast diffusion-controlled processes, the adsorption of reactants is a relatively fast step compared with succeeding electron transfer steps and can be considered in quasi-equilibrium. A knowledge of reactant adsorption behavior is necessary for interpretation of the mechanism of the reaction. Equilibrium adsorption studies are directed toward the evaluation of the surface concentration of reactants in relation to the electrode potential, the temperature, the activity of reactants, and other species in the bulk and the energy of adsorption as a function of the partial coverage θ. Study of the surface coverage by adsorbed intermediates can in some cases give additional information to the kinetic approach. Determination of adsorbed intermediates would indicate the path which the reaction follows.

The experimental methods available for determination of the surface concentration of adsorbed intermediates formed in charge-transfer processes have been reviewed recently by Gileadi and Conway.[90] They have been divided into two groups: (a) methods in which the charge needed to form or desorb the surface species is directly measured; and (b) methods in which the capacity at the interface (i.e., the sum of the ionic double-layer capacity and the adsorption pseudocapacity) is measured as a function of potential and integrated to obtain the charge. These methods will not be discussed further here. The experimental methods described below pertain to the determination of adsorption of uncharged species on solid electrodes.

(i) Electrochemical Methods

(a) General—The most accurate method of determining adsorption at electrodes is by measurement of the variation of the excess surface free energy (surface tension) with potential and with the activity of the relevant species in the bulk of the solution. This method has been employed extensively by the Russian School of electrochemists and by Grahame and Parsons and Bockris *et al.* for the study of adsorption on mercury electrodes. Unfortunately, this method is limited to liquid metals, since an accurate way of determining the excess surface free energy of solid electrodes has yet to be devised.

Grahame has used the differential double-layer capacity method extensively to determine surface coverage on mercury. The fractional surface coverage can only be obtained from capacity data in an unambiguous (thermodynamically rigorous) manner if the coordinates of the electrocapillary maximum (i.e., the potential at which the surface tension is a maximum, and the value of the surface tension at this potential) are known. Thus, the capacitance method can be applied quantitatively only to liquid-metal electrodes. Approximate methods, relying on nonthermodynamic considerations have been suggested.[91,92] The validity of one of these methods[92] has been questioned, however,[93] and the range of applicability of the other[91] is limited. A case of strong adsorption (ethylene on platinum) associated with essentially no change of capacity has recently been reported.[94,95]

Spurious frequency variations which may arise because of surface roughness, and complications because of contributions from adsorption pseudocapacitance make the interpretation of capacity data on solid electrodes even more qualitative.

An additional difficulty which arises in the interpretation of adsorption measurements on solid electrodes is the uncertainty with regard to the real surface area of the electrode. Various methods have been suggested for the determination of the roughness factor, (i.e., the ratio between real and apparent surface area) none of which is quite satisfactory.[90] Frumkin *et al.*[96] obtained the roughness factor by comparison of the capacity in the double-layer charging region (i.e., where no Faradaic process takes place) with that obtained on mercury at the same rational potential.* Brodd and

---

*The term "rational potential" was first introduced by Grahame[98] and refers to the potential relative to the potential of zero charge (p.z.c.) in the same system.

Hackermann[97] have conducted extensive studies in comparing this method with the B.E.T. gas-adsorption method using krypton and have found reasonable consistency between the two methods for most metals studied. The charge required to form a complete monolayer of oxygen or hydrogen has been used by various authors[99,100] as a measure of the real surface area of the electrode. However, the implicit assumption of a simple stoichiometry on the surface, that is, the assumption that one oxygen or hydrogen atom is adsorbed per metal atom at saturation, may be questioned. Furthermore it has recently been shown[101] that substantial amounts of hydrogen and even oxygen may be absorbed in platinum, so that care must be exercised in using hydrogen and oxygen charging methods for surface-area determination.

In conclusion it may be said that the various methods yield the right order of magnitude of the roughness factor and may be used as a guide for the estimation of the real surface area, but not as an accurate measure of it.

(b) *Coulometric methods*—These methods are applicable for determining fractional surface coverage when the adsorbed species undergo a single quantitative electrode process at 100% current efficiency. They are limited to conditions such that neither mass transport nor chemical reaction can act to produce more of the electrochemically active surface layer during the measurements.

(α) Galvanostatic charging curves—The procedure involves applying a constant current pulse at high current density to oxidize or reduce the adsorbed species. A potential arrest during which the adsorbed species is oxidized is usually observed. The corresponding charge (corrected for the charge consumed in the same region in the absence of adsorbate) is then compared with the charge associated with a monolayer, obtained either from theoretical considerations or experimentally determined by deposition of a monolayer of hydrogen or oxygen, and thus converted to fractional coverages. The fast galvanostatic charging method can be applied to systems for which a clearly distinguishable potential arrest is obtained in the transient curve. It is largely limited to the noble metals and restricted to systems in which the slow electron transfer step involves the substance whose coverage is required.

(β) Potential sweep—Fractional surface coverages have been determined from current–potential curves obtained using a single triangular potential sweep of very short duration, e.g., Breiter and Gilman have used a voltage sweep rate of 800 V/sec,[102] for the study of methanol adsorption. The area under the current peak, minus the area for a similar sweep in the absence of organic, is supposed to give the charge required for removal of the adsorbed species by anodic oxidation. The applicability of this potentiostatic technique is limited by several considerations:

1) The charge used for the anodic formation of an oxygen layer and for oxygen evolution during the sweep is assumed the same in the absence and presence of organic species. Hickling and Wilson reported that the oxygen overpotential at a platinum anode was shifted appreciably by the addition of oxidizable substances containing N or S to a phosphate buffer solution.[103] An elevation in the overpotential of 0.63 V was observed for addition of thiourea and of 0.41 V for urea. It was shown recently[33,104] that the potential of oxide formation is shifted by addition of hydrocarbons to the system. Thus, due to a "poisoning" effect of organic molecules, it is not expected that the above assumption is valid. For systems in which the organic species is completely oxidized during the sweep before the potential of oxide formation is reached, the method may not be subject to this criticism.

2) The capacitative current required to charge the ionic double layer is assumed to be the same in the presence and absence of organic species. The decrease in the double-layer capacity by addition of organic substances is from about 16 μF/cm^2 to 4–8 μF/cm^2 on Hg in aqueous solutions. Breiter found a change from about 60 μF/cm^2 to 25 μF/cm^2 for adsorption of methanol from perchloric acid using platinum anodes.[105] In the case of methanol[102] the change in ionic double-layer capacity will introduce a negligible error (the maximum current density during the transient was 1.2 A/cm^2 while the error $\Delta C(dv/dt)$ amounts only to 0.03 A/cm^2), but for other organic substances which are not so readily oxidized this may constitute a major error.

3) The anodic sweep should be sufficiently rapid to eliminate diffusion of organic molecules to the electrode during the oxidation process. This assumption is easily tested by experiment and can be readily satisfied.

4) The high sweep rates necessary to avoid diffusion of species to the electrode may result in incomplete oxidation of the adsorbed organic material. For example, the studies of Greene and Leonard on metal corrosion show that passive current densities are markedly altered by traverse direction, traverse rate, and starting potential.[106] They also observed that regions of passivity were missed at increased sweep rates.

5) Electrostatic desorption, i.e., the decrease of coverage with increasing potential due to the electrostatic interaction between water dipoles and the field in the double layer, may give rise to too low values of the coverage.

It appears that the fast potential sweep method may be used for measurement of coverage by organic in certain simple cases, but a test should be made to verify the implicit assumptions in each case.

(γ) Galvanostatic hydrogen deposition—This method is based on the relation

$$\theta_{\text{org}} = \frac{(Q_H^S - Q_H)}{Q_H^S} \qquad (14)$$

where Q_H^S is the number of coulombs/cm^2 required to form a monolayer of hydrogen on the electrode and Q_H is the charge required to fill the sites available for hydrogen adsorption in the presence of organic species, as obtained from charging curves in the absence and presence of the organic adsorbate, respectively.[107–109] Assumptions made in this method are that the organic species adsorbed on the surface is unaffected by the cathodic charging process, the heat of adsorption of hydrogen is not affected by the presence of adsorbed organic molecules, and any site accessible to hydrogen is also accessible to the organic molecule.

If the charging process is initiated at any potential above the rest potential, which is necessary to determine θ as a function of potential, the organic species would certainly be oxidized, the extent of oxidation being dependent on the starting potential and the charging current. Below the rest potential, the organic species can be readily reduced, e.g., see the extensive literature on polarographic reduction of organic compounds. The question of whether the same sites accessible to hydrogen atoms are also accessible to the much larger organic molecules depends on the arrangement of the species on the surface. If the organic molecules could be packed tightly together on a smooth surface, it is possible that the accessible sites are the same, accounting for the difference in the number of sites required per molecule. However, the metal surface may be quite rough providing regions accessible to hydrogen but not to the larger organic molecules. Also it is probable that the organic species cannot be packed tightly on the surface because of steric hindrance, thus leaving free sites between molecules which are accessible to hydrogen but not to the other organic species. This method is further questioned on the basis of recent results obtained by Schuldiner and Warner in their study of hydrogen and oxygen adsorption and absorption at platinum electrodes.[101,109] They observed that on galvanostatic charging, several monolayers of hydrogen or oxygen could be readily absorbed within the platinum, occurring to a greater extent with hydrogen than with oxygen. A detailed discussion of the last two methods and the approximations involved in their use has recently been given.[110]

(δ) Chronopotentiometry—A variation of the galvanostatic charging method has been suggested by Laitinen, when mass transfer is the only process involved.[111] The total transition time due to adsorbed and solution phases is determined and the diffusion contribution is extrapolated out. This method has not been used and would not seem to offer any advantages over the charging method where diffusion control is not involved.

(c) Differential double layer capacity methods—The capacity of the double layer is altered by any chemical or physical change which alters the effective dielectric constant or the effective thickness of the double layer. The adsorption of an organic species and simultaneous desorption of water molecules is associated with a change of dielectric properties of the metal–solution interface. For a rigorous determination of coverage from capacity measurements it is necessary to have the coordinates of the electrocapillary maximum (i.e., the potential of zero charge and the value of the excess interfacial energy at this potential). Several approximate methods have been suggested to evaluate coverage from capacity data on solid electrodes where the electrocapillary curve cannot be measured. These have been reviewed recently.[93] The equation suggested by Frumkin[91] is very simple and constitutes a good approximation in certain potential regions. The charge on the metal at a coverage θ may be written according to Frumkin as

$$q_\theta = q_{\theta=0}(1 - \theta) + q_{\theta=1}\theta \qquad (15)$$

where $q_{\theta=0}$ and $q_{\theta=1}$ are the values of the charge on the bare metal and on the completely covered metal at the same potential. Differentiating equation (15) with respect to potential one has

$$C_\theta = C_{\theta=0}(1 - \theta) + C_{\theta=1}\theta + (q_{\theta=1} - q_{\theta=0})\frac{d\theta}{dV} \qquad (16)$$

In the range where the coverage varies little with potential it is possible to neglect the last term in equation (16) and one obtains

$$\theta = \frac{C_\theta - C_{\theta=0}}{C_{\theta=1} - C_{\theta=0}} \qquad (17)$$

where the subscripts for the capacities are the same as for the charges in equation (15). Equation (17) will be approximately applicable at potentials around the potential of maximum adsorption, i.e., from the point of zero charge a few hundred millivolts in the cathodic direction.[112]

Capacity measurements are very useful in determining the coverage by adsorbed intermediates formed in fast charge transfer processes, which give rise to an appreciable adsorption pseudo-capacitance. The experimental methods which can be employed have been reviewed recently,[90] and will not be discussed here further.

(d) Rotating disk electrode with a ring—The rotating disk electrode was developed and has been used to an increasing extent by the Russian School of Electrochemistry.[114] The use of a rotating electrode with a ring is essentially different from the methods discussed previously since the aim is not to determine species adsorbed on the electrode surface but to detect intermediate species which are easily desorbed. A rotating platinum disk electrode with a platinum ring has been successfully used to detect hydrogen peroxide at the oxygen electrode.[113] The principle is that stable intermediates are desorbed from the platinum disk and diffuse to the ring encircling the disk where they may be detected by electro-chemical oxidation or reduction. For an oxidation reaction of the type $A \xrightarrow{n_A e} B \xrightarrow{n_B e} C$, where B is an intermediate in the overall reaction, the following relation has been given:

$$i_{ring} = \frac{n_B}{n_A} \frac{N i_{disk}}{1 + k\delta/D} \tag{18}$$

where i_{ring} is the current obtained in the electrode process involving the intermediate B at the ring, n_A is the number of electrons involved in oxidation of A to B at the disk, n_B is the number of electrons involved in the reduction of intermediate B at the ring, N is a factor between 0 and 1, i_{disk} is the rate of formation of the intermediate at the disk, k is the rate constant for further oxidation of B to products C at the disk, δ is the diffusion layer thickness, and D is the diffusion coefficient of B. The fraction of intermediate species which is desorbed will depend on the metal–intermediate bond strength, giving this method a somewhat qualitative aspect. Information on intermediates would not be obtained with the ring if the intermediates are strongly adsorbed. In reactions where several products are formed, possibly through formation of stable free radicals, this method may offer a valuable approach. For example, in the oxidation of ethylene on gold or palladium, several products are formed, particularly compounds containing three carbon atoms.[70]

(ii) Direct Methods

Several methods are available for following the equilibrium adsorption of reactants and other species at anodes. The more successful methods, based on the removal of adsorbate from solu-tion onto the electrode, are now fairly well developed.

(a) Radiotracer methods

(α) Isotope-labeling detection in solution—The change of specific activity of electrolyte due to removal of labeled species from solution by equilibrium adsorption on the electrode has been studied primarily by the Russian School.[115–118] This method is limited to low concentrations (below approximately 10^{-6} mole/liter), since the change in concentration because of adsorption is negligible at higher concentrations of adsorbate. Also this approach cannot be used at potentials where reaction of the labeled species occurs, limiting its application to low temperatures where oxidation of the organic species is negligible.

(β) Isotope-labeling detection on the electrode—The detection of labeled ions or molecules on electrodes was made initially by Hevesy and Weiss, who removed metal sheets from the electrolyte to determine the activity of the adsorbed material.[119] Erbacher[120] continued this type of study and more recently Hackermann and Stevens[121] have employed this method, removing the electrode and washing it prior to activity determinations. In this procedure a subtraction of two comparable specific activities is avoided. However, the potential control and thermodynamic equilibrium are lost and only highly irreversible and potential-independent adsorption can be studied.

The *in situ* detection of labeled species adsorbed on an electrode remaining in contact with the electrolyte raises several experimental difficulties because of shielding and background count in the solution. The procedure of making the window of a Geiger–Müller counting tube the electrode on which the radio-labeled species is adsorbed and directly counted, was first employed by Joliot,[122] in his study of the electrodeposition of Po. This method was developed independently by several investigators in 1949 and 1950.[123–127] The background count of the solution was determined as a function of the distance of the counter tube window from the surface of the solution and the difference in count when the window made contact with the solution was attributed to adsorption of the labeled species. In 1959, Cook reported results obtained for adsorption of radio-labeled steric acid and stearyl alcohol from n-hexadecane using mica windows specially coated with iron and gold.[128]

The development of suitable apparatus for adsorption studies by this method was reported by several authors in 1958–59.[128–131] Blomgren and Bockris[132] reported the further development of the method of Anionsson,[125] in 1960. Schwabe has continued studies of adsorption, particularly of radio-labeled ions, e.g., SO_4^{2-}, Cl^-, I^-, ClO_4^- and CN^-.[133–135] The procedure used by Blomgren and Bockris has been extended to studies of the adsorption of ethylene on platinized gold foils[136] (gold foil was used over the counter window since most common metals cannot be obtained in a form thin enough for the weak β-radiation from C^{14} to pass through and be counted), thiourea,[137] benzene, naphthalene, phenanthrene, and cyclohexene[138] on gold, and further studies of ethylene, benzene, and other hydrocarbons on platinized gold are in progress.[94]

Another method in which an electrode in the form of a tape is passed through the solution containing organic adsorbate and then withdrawn through a narrow slit into a proportional counter has been successfully developed by Bockris, Green, and Swinkels.[139–141] The absolute accuracy of this method is limited mainly by the error in measurement of the true surface area, and was reported to be $\pm 65\%$.[141] This error is common to all adsorption measurements on solid electrodes. The reproducibility of the method was estimated by Bockris and co-workers to be 25%. The design of the cell is important since any IR drop in the region of the slit can lead to a different potential than that in the bulk of the solution. This could result in significant errors in the measurement of activity of the adsorbed species outside the cell. An advantage of this method is that thin metal foils are not necessary and a wider range of metals can be examined.

(b) *Spectrophotometric methods*—A direct spectrophotometric method for determination of adsorption at solid metals was developed by Conway and co-workers,[142] for substances exhibiting adsorption in the UV or visible spectrum. The change in adsorption of the solution (up to 70% of the initial value) due to removal of adsorbate by adsorption on large (700 cm^2) gauze electrodes was determined with a Beckman DK2 spectrophotometer. This method is particularly suitable for organic molecules containing conjugated double bonds and was used for the study of adsorption of pyridine, quinoline, and acridine from aqueous solutions and polyvinyl pyridine from methanol solutions on copper, nickel, and silver electrodes. The method is limited to a range of potentials in which no oxidation or reduction of the organic species occurs.

Infrared adsorption spectroscopy has greatly aided the study of heterogeneous catalysis and is in principle applicable to electrode adsorption studies. The major problems involved are, (a) To obtain an adequate spectrogram several monolayers of adsorbate must be traversed by the beam of infrared light. This requires either many reflections from an adsorbing plate or transmission through a high surface-area sample, with sufficient intensity retained in the infrared beam to permit spectral study. (b) Interfering absorption of infrared radiation by aqueous or alcoholic liquid phase in contact with the adsorbed monolayer must be minimized or eliminated. (c) The adsorbing film (e.g., platinum) must be electrically conducting to allow control of potential.

A study of the feasibility of using infrared methods was conducted by the American Oil Company.[143] Their results indicated that infrared studies of electrode surfaces by transmission or direct-reflection methods appear less promising than the internal-reflection technique. Even under favorable conditions, multiple internal-reflection studies (using germanium and silicon) proved

very difficult. The study of C—H, C=O, C⟨$_O^O$ -groups for mono-

layers of acids down to about C_5 could be made but a monolayer of methanol could not be detected. It was concluded that formidable experimental difficulties stood in the way of a general application of this method, and that application of the method was possible but considerably more development was required.

The possibility of applying the electron-spin-resonance technique to the study of free-radical intermediates in organic electrode reactions was also investigated by the American Oil Company.[143] Studies using stable free radicals (e.g., from p-amino phenol) gave an indication of optimum conditions for detection. However, application of these conditions to the search for anodically generated radicals from benzyl alcohol, t-butanol, and p-nitro benzyl alcohols did not produce observable ESR signals. It is felt that further attempts in this direction will be worth while.

(c) Ellipsometry—The application of ellipsometry, the measurement of the change of polarization state of elliptically polarized light upon reflection from a surface[144] to the study of calomel film formation on mercury[145] and oxygen on platinum electrodes[146] has been recently made. The study of adsorption of organic species in the gas phase has also been made with ellipsometry. These investigations suggest the possibility of determining fractional surface coverage of organic species from solution by ellipsometric methods. Such a study has not at present been reported. One problem which might be encountered is whether the refractive index of the adsorbed organic species would be sufficiently different from that of the solution to give measurable results.

4. Evaluation of Kinetic Parameters

The primary aim of any electrode kinetic study should be to establish the reaction mechanism by experimental evaluation of the kinetic parameters. The usual mechanistically significant kinetic parameters (e.g., reaction orders, apparent energies of activation, specific rate constant) are useful in electrode kinetics. Since a hetereogeneous process is involved, the extent of adsorption of reactants, product, intermediates, and poisons or inhibitor is also of great importance. The peculiar feature of electrode kinetics is the possibility to control the current density or the potential (i.e., the rate or the effective standard free energy of activation respectively) at constant temperature and composition. The Tafel slope $b = (dV/d \log i)_{\mu_j, T}$ can further characterize the reaction mechanism and in particular the position and nature of the rate-determining step in the reaction sequence. Several mechanisms can, however, often lead to the same numerical value of b, particularly at intermediate values of the coverage when Temkin conditions prevail.[90] Thus, the Tafel slope can be regarded as an important, but not in itself sufficient, parameter in the elucidation of electrochemical reaction mechanisms.

The stoichiometric number, defined as the number of times the rate-determining step has to occur for every complete act of the overall reaction, was first used by Horiuti et al.[147,148] for the study of electrode reactions. It can only be measured properly for highly reversible reactions (e.g., the hydrogen evolution reaction on some catalytic metals, some metal dissolution reactions), and hence it finds no application in the study of the anodic oxidation of most organic compounds.

Electrode reactions are usually quasi-zero order with respect to reactants, either because the solvent itself serves as the reactant (e.g., during hydrogen or oxygen evolution in dilute aqueous media) or because a constant supply of reactant is maintained during the experiment. This has led to the unfortunate tendency to neglect reaction order studies in electrochemistry, particularly in the field of anodic oxidation of organic compounds relevant to fuel cell technology. The importance of determination of reaction order in electrode kinetics has been emphasized by Vetter[149] and the measurement of this parameter should constitute an essential part of any kinetic study aimed at the evaluation of the mechanism of the reaction. The experimental methods by which the kinetic parameters may be determined will be discussed briefly below.

(i) *Steady-State Methods*

The classical approach to mechanism studies is that in which the rate of reaction is measured directly in terms of current. The normal steady-state current–potential behavior is evaluated with standardized conditions of electrode preparation and solution purity. Standardized times of measurement are essential for anodic organic reactions owing to the long time variation of current or potential with solid metals.* The galvanostatic method, i.e., measurement of electrode potential at constant current, was the only method feasible for such studies until the innovation of the potentiostat by Hickling.[18] With a potentiostat, the current in an electrochemical cell is regulated from a counter electrode to a measuring electrode in such a manner that the potential difference between the sensing point of a capillary in contact with a reference electrode and the measuring electrode equals the assigned potential of the instrument. The operating principles of a number of such instruments have been described in the literature.[87] The potentiostatic method, i.e., measurement of current at constant potential, is usually preferred for kinetic studies, and is essential in cases where

---

*The standardized times of measurement are not arbitrarily chosen. For example, in a study of the anodic oxidation of ethylene in alkaline solution, the dependence of the Tafel slope on the degree of approach to steady state was determined.[69] The current–potential curves determined potentiostatically at 5 min intervals for approach to steady state gave Tafel slopes of about 300 mV. These slopes decreased with increasing length of time for steady-state current to be established, and for 60 min and times greater than 60 min, up to several hours, a slope of about 150 mV was obtained. Since a slope of 150 mV, i.e., $2 RT/F$, is a rational slope which is significant for interpretation in electrode kinetics, and 300 mV is not a rational slope, the use of longer times for the establishment of steady-state conditions is justified.

a passivation or quasi-passivation process occurs, characterized by a decrease of current with increasing potential (i.e., a region of negative reaction resistance).

The kinetic parameters are most easily obtained potentiostatically, as indicated previously, by measuring the change in current with variation of a single component, maintaining all other variables constant. A problem arises in the determination of stoichiometric factors in relating the surface concentrations to bulk concentrations of reactant ions. In general this is not done and reaction orders are expressed in terms of the bulk concentration.

The application of steady-state methods to anodic organic oxidation reactions has been criticized because of the time variation observed which can lead to considerable irreproducibility. While the time variation remains a problem, if standardized procedures and electrode preparations are employed, good reproduction of data (within 10%) can be obtained.[32,33] Steady-state methods are limited to the study of reactions involving a high energy of activation where diffusion control is not important. All of the anodic oxidations of organic species reported up to now have satisfied this requirement. The experimental details for steady-state methods have been well established in the literature and require no further elaboration.

(ii) Nonstationary Methods

(a) Relaxation methods—The application of relaxation methods to electrode kinetics has been recently reviewed by Delahay[150,151] and by Reinmuth.[152] The technique is to examine the behavior of an electrode during a relatively short time following a modification of the equilibrium conditions (usually by variation of the current or potential, although nonelectrical perturbances have also been employed).[153,154] Relaxation methods can be classified as small amplitude or large amplitude techniques and in most cases studies are made by (a) the potential-step method, (b) the current-step method, or (c) electrolysis with superimposed alternating potential. The development of these methods over the last 20 years is one of the major advances in the study of fast electrode reactions.

The classical (steady state) method of determination of the kinetic parameters is applicable to reactions sufficiently slow, corresponding to a rate constant of $k < 10^{-4}$ to 10^{-5} cm/sec. This is not an absolute limit but relaxation methods are preferable for the study of reactions with $k > 10^{-5}$ cm/sec. The method of electrolysis with superimposed AC potential is the most precise because the kinetic parameters are calculated starting with bridge measurements rather than oscillographic recordings. This method is applicable in the range $10^{-5} < k < 1$ cm/sec. The potential-sweep and current-step methods are applicable for values of k less than about 1 cm/sec. (It can be shown on the basis of calculation of the maximum number of collisions that k cannot exceed the limit 10^3 to 10^4 cm/sec.)[151] The limitations in applicability of most relaxation techniques to kinetic studies at high frequencies arise because of the double-layer capacitance. This has been eliminated to some extent by the development of "second-order relaxation techniques" such as the double-pulse method.[155] Since the electrode reactions with which we are concerned, i.e., anodic organic oxidations, all have rate constants well below the range where diffusion control becomes appreciable, the small-amplitude relaxation techniques will not be further discussed.

The widespread use of large-amplitude relaxation techniques in the investigations of anodic organic oxidations, requires further comment on the value of these methods. Reinmuth[152] divided these techniques into three classes based on the types of applications: quantitative kinetic studies, qualitative kinetic studies, and analytical studies. We are not concerned here with the analytical applications. For studies in kinetics, controlled-potential techniques, particularly linear-potential scan, in either single sweep or in cycles, and to some extent chronopotentiometry, have been primarily employed. Chronopotentiometry has been successfully utilized in the study of transient reactions, e.g., the reaction of CO with platinum oxide[109,156] or the reaction of oxalic acid with platinum oxide,[157,158] and the study of simple charge-transfer reactions with linear diffusion (cf. Refs. 159–161). However, since the general application of chronopotentiometry is severely limited for the study of anodic organic oxidations, as commented previously, this technique will not be further discussed. The quantitative analysis of data obtained by linear potential scan techniques is complicated because the form of theoretical results, even for the simplest cases, requires the use of computers and consequently very little quantitative kinetic information has been obtained. This method is popular since qualitative detection of intermediates or adsorbed species simply requires counting peaks on the polarogram. However, this is not the most profitable approach to establish the mechanism of a reaction. This method of linear potential scan is now further discussed in detail.

(b) *Linear potential scan techniques*—The application of a triangular potential sweep was made in polarography in 1948,[162] however, not as applied from a potentiostat. The potentiostatic method of triangular potential sweep was developed and applied to the study of hydrogen and oxygen adsorption by Will and Knorr.[100,163] The application of this method to organic electrode processes has been recently reviewed by Juliard[164] and by Vielstich.[165] The method was indicated to give only qualitative information but to be a valuable supplement to the usual methods of kinetic experimental procedure by which data could be obtained in a relatively short time, with reproducibility resulting from a decreased effect of solution impurity and a supposed reproducible surface and easy adoption to various situations, i.e., separation of equilibrium and nonequilibrium processes and the study of mixed electrode processes. However, a number of problems are inherent in the interpretations of results.

Cycling of an electrode between fixed potentials may lead to a roughening of the metal surface, giving an increased surface area which depends on the values of the fixed potentials. On increasing the anodic-potential limit of the cycle to potentials above about 1.6 V, a bright platinum electrode becomes noticeably black, exhibiting properties similar to those obtained on platinization. Hoare has reported similar effects by the application of alternating current to platinum electrodes.[166]

This picture is further complicated by the investigations of Schuldiner and Warner,[109] who found that hydrogen and oxygen are readily absorbed into the platinum surface. Thus by cycling a platinum electrode into hydrogen or oxygen evolution, the surface structure of the metal would be changed. Thus the time necessary to obtain a quasi-steady-state surface area will depend on the potential limits of the cycle. This has not, however, been defined in cyclic voltametric studies and the assumed reproducible surfaces have not been satisfactorily demonstrated. The work of Will and

Knorr showed that the polarograms obtained by the triangular potential-sweep method were distinctly dependent on the sweep rate and the potential limits over which the cycle was conducted. Will and Knorr have stated: "The increase of the voltage rate has not only an effect upon the height of the H_{ads} and O_{ads} peaks but changes also the shape of the curve."[100] The positions of the current peaks resulting from electrode reactions were found to shift with sweep rate and the measured currents were found to be related to the sweep rate by

$$i = C \left(\frac{dV}{dt} \right)^k \qquad \text{with } k \leqslant 1 \qquad (19)$$

where C has the dimensions of capacitance.

Srinivasan and Gileadi[167] have analyzed the potential sweep method from the theoretical viewpoint and derived equations relating the shift in the adsorption peaks and the changes in the form of the V–i curves with sweep rate to the specific rate constants of the reaction involved.

In a comparative study of potentiostatic anodic polarization methods, Greene and Leonard found that the anodic polarization characteristics were strongly controlled by the experimental procedure employed.[106] They concluded that while potential sweep techniques yielded the most reproducible results, they were not necessarily the most accurate and that measurements should be conducted as slowly as possible, even though only qualitative data are required.

The theory for cyclic voltametric methods has not been worked out quantitatively. In his discussion, Delahay indicated that the initial concentration of reacting species will vary from one cycle to another and the initial conditions for the boundary value problem cannot be easily stated.[86] As a result the experimental peak currents should not necessarily be in agreement with derived theoretical values. Delahay concluded that compared with conventional voltametry, voltametry with continuously changing potential has several disadvantages which limit its usefulness.[86]

If the capacity is a function of potential, the measured currents will contain a contribution from capacitative currents which will depend on the sweep rate and may become appreciable at higher sweep rates. The fractional surface coverage of the electrode as determined from the potential-sweep method would be expected to differ from that pertaining to steady-state equilibrium adsorption. This difference has in fact been observed in adsorption studies of formic acid on platinum.[39] Conway has recently commented on the significance of measurements by the triangular-voltage-sweep method applied to adsorption reactions,[168] and this has been subsequently expanded.[169] The derived equations indicate a complex variation of the capacity with sweep rate, involving several terms in dV/dt, and the adsorption pseudocapacities measured at significant sweep rates are then difficult to interpret. Conway concluded: "The steady, slow, point-by-point potentiostatic examination of a current potential relationship is to be preferred for evaluating kinetic behavior, and the direct differential galvanostatic method is to be preferred for dealing quantitatively with charging processes, since the current density is kept constant and is measured, and the resulting equations for the variation of the fractional coverage and the adsorption pseudocapacity with time and potential are much simpler." Since the charge associated with the change of coverage between two potentials will be a function of sweep rate in cyclic voltametric experiments, the use of this method for estimating coverage by intermediates in complex reactions is questioned except for relatively reversible reactions. Conway also noted that the sweep rate may have an indirect effect on kinetic currents which are partially determined by coverage of intermediates produced in the reaction or by coverage of species entering into competing reactions.

It is concluded that while this method is certainly of value as a supplementary experimental method for preliminary electrode kinetic studies, at the present stage of development of its theory, it is not the best approach or even a good approach for the evaluation of kinetic parameters.

III. ADSORPTION

1. General

The electrosorption of organic compounds as determined by the electrocapillary method and from measurements of differential capacity, has been reviewed in Volume III of this series by Frumkin and Damaskin.[93] Their discussion applies primarily to adsorption on mercury electrodes on which the bulk of electrosorption studies have been carried out. The behavior of intermediates formed in a charge-transfer process, which give rise to an adsorption pseudo-capacitance, has been extensively discussed by Gileadi and Con-way.[90] While much of the data in this area have been devoted to hydrogen and oxygen evolution and dissolution reactions, some of the recent studies have treated organic oxidation. The area in which we are primarily interested, i.e., the electrosorption of organic molecules which serve as reactants for oxidation reactions on solid electrodes, has until recently received little attention because of the limitations imposed on the experimental methods. Since the experimental methods have already been reviewed (cf. p. 55), we present here a brief summary of the available information for electrosorption of organic compounds at solid electrodes.

2. Special Aspects of Adsorption on Solid Electrodes

Before presenting the experimental data for adsorption of organic compounds, it will be instructive to examine the fundamental differences between electrosorption, i.e., adsorption on electrodes and adsorption from the gas phase.

1) The major difference between gas phase adsorption and electrosorption is that in the former case adsorption occurs on a bare surface, while in the latter case the metal substrate is solvated, i.e., covered with an adsorbed layer of solvent molecules. Thus it is evident that adsorption on electrodes is a replacement reaction. The observed standard free energy, enthalpy and entropy of adsorption can therefore only be related to the type of interaction between the adsorbate and the electrode if the corresponding thermo-dynamic quantities for the solvent are known and the number of solvent molecules replaced by each adsorbed organic molecule can be estimated.

2) Since the effective standard free energy of adsorption is the difference between the corresponding quantities for solute and solvent, the solvent is expected to have a leveling effect, making the differences between different sites on the surface, and even between different metals less pronounced. In the few instances where the rate of change of the apparent standard free energy of electro-sorption with coverage has been measured, it was found to be much less than the corresponding values in gas phase adsorption. For

example, this was found to be on the order of 1–3 kcal for the adsorption of several butyl, phenyl, and naphthyl compounds on mercury[170] and for the adsorption of n-decylamine[140] and naphthalene[141] on several solid metals.

3) The extent of adsorption is dependent on the solubility of the organic species. In general, the lower the solubility, the higher will be the adsorbability. A systematic study of the adsorption of butyl, phenyl, and naphthyl compounds at mercury electrodes indicated a linear relationship between the standard free energy of solvation and the standard free energy of electrosorption within each group.[170] In a recent comparative study of the adsorption of butanol and phenol on mercury from aqueous and methanolic solutions, adsorption from aqueous solutions was found to be much more extensive at a given bulk concentration.[171] This is clearly because of the lower solubility of both alcohols in water, since on the basis of competition with solvent only, adsorption from methanol would be favored.

4) Perhaps the most important feature which distinguishes electrosorption from gas phase adsorption is the presence of the metal–solution potential difference as an additional degree of freedom of the system, which can be controlled and varied externally. The potential region over which adsorption occurs and the manner in which adsorption depends on potential will greatly influence the oxidation behavior of a given compound.

5) The use of solid electrodes gives rise to several problems which must be considered; the electrocapillary method cannot be applied and the differential capacity method is rather limited. The real surface area of solid electrodes is not easily established and adsorption on different crystal faces, on grain boundaries, or on dislocation sites, may play an important role in determining adsorption.

3. Adsorption as a Replacement Reaction

The most significant feature of electrosorption as compared with gas phase adsorption is that the former is a replacement reaction while the latter is not. Bockris, Green, and Swinkels have shown[140] that for an organic species R adsorbing on the surface and replacing n water molecules, the adsorption process may be represented as

$$R_{soln} + nH_2O_{ad} \rightleftharpoons R_{ad} + nH_2O_{soln} \qquad (20)$$

where the subscripts soln and ad refer to the species in solution and adsorbed on the surface, respectively.

The chemical potentials of the species involved may be written as

$$\mu_{R_{soln}} = \mu^0_{R_{soln}} + RT \ln X_{R_{soln}} \doteq \mu^0_{R_{soln}} + RT \ln \left(\frac{C_{R_{soln}}}{55.4}\right) \quad (21)$$

$$\mu_{H_2O_{ad}} = \mu^0_{H_2O_{ad}} + RT \ln X_{H_2O_{ad}} \quad (22)$$

$$\mu_{R_{ad}} = \mu^0_{R_{ad}} + RT \ln X_{R_{ad}} \quad (23)$$

$$\mu_{H_2O_{soln}} = \mu^0_{H_2O_{soln}} + RT \ln X_{H_2O_{soln}} \doteq \mu^0_{H_2O_{soln}} \quad (24)$$

The condition for adsorption equilibrium is

$$\mu_{R_{soln}} + n\mu_{H_2O_{ad}} = \mu_{R_{ad}} + n\mu_{H_2O_{soln}} \quad (25)$$

The mole fractions on the surface are given by

$$X_{R_{ad}} = \frac{\theta}{\theta + n(1 - \theta)} \quad (26)$$

$$X_{H_2O_{ad}} = \frac{n(1 - \theta)}{\theta + n(1 - \theta)} \quad (27)$$

where θ is the fractional surface coverage by the organic species R. This leads to the isotherm

$$\frac{\theta}{(1 - \theta)^n} \frac{[\theta + n(1 - \theta)]^{n-1}}{n^n} = \frac{K_{C_R}}{55.4} \quad (28)$$

where

$$K = \exp\left(\frac{-\Delta G^0_{ad}}{RT}\right) \quad (29)$$

and

$$\Delta G^0_{ad} = (\mu^0_{R_{ad}} - \mu^0_{R_{soln}}) - n(\mu^0_{H_2O_{ad}} - \mu^0_{H_2O_{soln}}) \quad (30)$$

The standard free energy of adsorption is thus the difference between the standard free energies of adsorption for the organic and for n water molecules. The same statement may be applied to the standard enthalpy and entropy of adsorption on electrodes.

Close examination of the energetic quantities involved in adsorption from solution shows[140] that the standard free energy of adsorption may be written as

$$\Delta G_{ad}^0 = RT \ln\left(\frac{c_{R_{sat}}}{55.4}\right) - RT \ln\left(\frac{P_R^0}{P_{H_2O}^0}\right) + \Delta G_V^0(R) - n\,\Delta G_V^0(H_2O) \quad (31)$$

where $c_{R_{sat}}$ is the concentration of R in a saturated solution, P_R^0 and $P_{H_2O}^0$ are the vapor pressures of pure organic and water at a temperature T, and ΔG_V^0 refers to adsorption from the gas phase.

The importance of the solubility and the relative vapor pressure of solvent and solute is thus brought out as factors influencing the extent of adsorption. It may also be noted that for the comparison of adsorbabilities of different solutes or the same solute in different solvents, a concentration corresponding to saturation may best be chosen as the standard state.[94]

4. Potential Dependence of Adsorption

Typical plots of the variation of the extent of adsorption with coverage are shown in Figures 2 and 3; the first figure refers to the

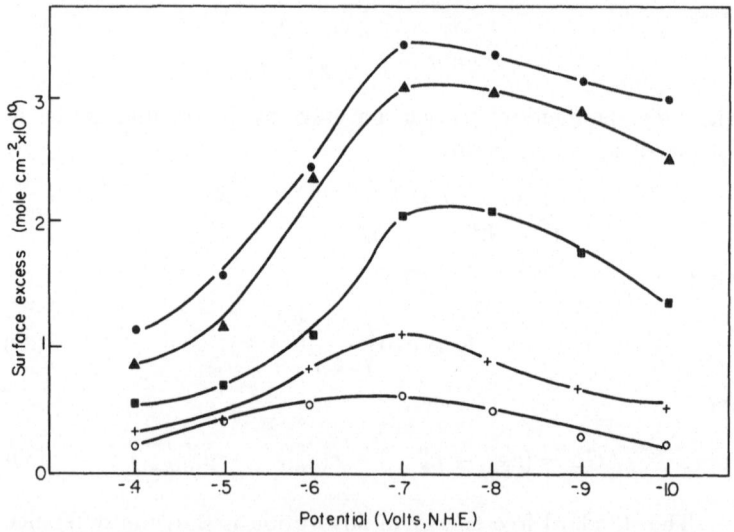

Figure 2. Coverage–potential relationships in the adsorption of n-decylamine on nickel from $1\,N$ NaClO$_4$: ($\bullet$) $7.5 \times 10^{-5}\,M$, ($\blacktriangle$) $5.0 \times 10^{-5}\,M$, ($\blacksquare$) $2.5 \times 10^{-5}\,M$, (+) $1.0 \times 10^{-5}\,M$, ($\bigcirc$) $0.5 \times 10^{-5}\,M$ [from Bockris and Swinkels, J. Electrochem. Soc. **111** (1964) 736].

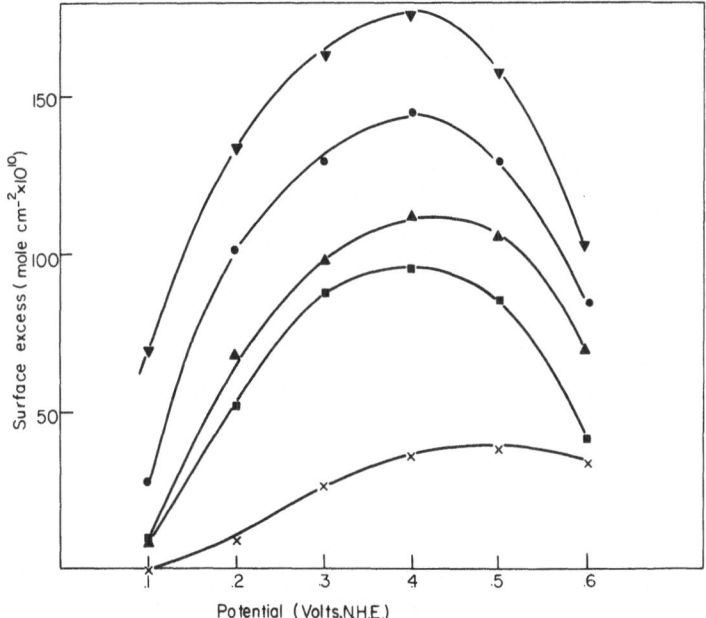

Figure 3. Coverage–potential relationship in the adsorption of ethylene on platinum from $1 N$ H_2SO_4: (▼) $1.7 \times 10^{-5} M$, (●) $9.1 \times 10^{-6} M$, (▲) $4.6 \times 10^{-6} M$, (■) $4 \times 10^{-6} M$, (X) $2.1 \times 10^{-6} M$ [from Gileadi, Rubin, and Bockris, *J. Phys. Chem.* Vol. 69, p. 3335 (1965)].

adsorption of n-decylamine on Ni[141] and the second to the adsorption of ethylene on Pt.[94] Similar results were obtained by Bockris and co-workers for the adsorption of n-decylamine[141] and naphthalene[140] on a number of metals (Ni, Fe, Cu, Pt, Pb) and in the adsorption of a number of butyl derivatives on mercury.[170]

The region of potential in which adsorption occurs is different for different metals. It depends mainly on the potential of zero charge of the metal and to a lesser extent on the specific interaction between the adsorbed water molecules and the metal.

The "bell shaped" form of the θ–V plot has been interpreted by Bockris and co-workers on the basis of the competition between organic and water molecules for surface sites and the field dependence of the standard free energy of adsorption of water,[94,112,140,141] as shown in the following section.

Consider first the water molecules as symmetrical dipoles of moment μ in the field X of the double layer, and assume that the free energy of adsorption of organic is entirely independent of field. The energy of interaction of dipoles with the field will be $\pm \mu X$ (the sign depends on orientation) and will become zero at the p.z.c. where the field is zero under the assumed conditions. Since the effective free energy of adsorption depends on the difference between the values for organic and water, maximum adsorption will occur at the p.z.c.

In practice the potential of maximum adsorption does not coincide with the p.z.c. and a somewhat more complex model should be considered.

Let N_1 and N_2 be the number of water molecules in the two possible orientations, and let the corresponding energies be E_1 and E_2. Then,

$$E_1 = E_1^c + \mu X - RcE \tag{32}$$

and

$$E_2 = E_2^c - \mu X + RcE \tag{33}$$

where E_1^c and E_2^c are the non-field dependent interactions of water in its two positions with the metal surface, c is a coordination number on the surface, E is the energy of interaction between dipoles, and R is given by

$$R = \frac{N_1 - N_2}{N_T} \tag{34}$$

where $N_T = N_1 + N_2$.

The average energy of a water molecule on the surface is

$$\bar{E} = \frac{N_1 E_1 + N_2 E_2}{N_T} \tag{35}$$

The potential E_{max} at which the average energy is a maximum can be calculated from the condition

$$\left(\frac{\partial \bar{E}}{\partial X}\right)_{N_T} = 0 \tag{36}$$

Thus

$$\left(\frac{\partial \bar{E}}{\partial X}\right)_{N_T} = \frac{1}{N_T}\left(N_1\frac{\partial E_1}{\partial X} + E_1\frac{\partial N_1}{\partial X} + N_2\frac{\partial E_2}{\partial X} + E_2\frac{\partial N_2}{\partial X}\right) = 0 \tag{37}$$

Now

$$\frac{\partial E_1}{\partial X} = -\frac{\partial E_2}{\partial X} \quad \text{and} \quad \frac{\partial N_1}{\partial X} = -\frac{\partial N_2}{\partial X} \tag{38}$$

hence

$$(N_1 - N_2)\frac{\partial E_1}{\partial X} + (E_1 - E_2)\frac{\partial N_1}{\partial X} = 0 \tag{39}$$

Equation (39) has an infinite number of solutions of which only the solution

$$N_1 = N_2 \qquad E_1 = E_2 \tag{40}$$

is physically meaningful. Thus if orientation 1 is stabilized by increasing field, one has $\partial E_1/\partial X < 0$ and also $\partial N_1/\partial X > 0$ and vice versa. In the region where position 1 is more stable, $(E_1 - E_2) < 0$ and there will be more species oriented in this direction, i.e., $(N_1 - N_2) > 0$. Thus the two products in equation (39) always have the same sign and they must therefore both be zero to fulfill equation (39). Since the two differential coefficients are not zero, the condition in equation (40) must hold.

Substituting for E_1 and E_2, with $R = 0$ one has

$$E_1^c + \mu X = E_2^c - \mu X \tag{41}$$

or

$$-\Delta E^c = E_1^c - E_2^c = -2\mu X \tag{42}$$

The field can be related to the charge per unit area q_m on the metal as

$$X = \frac{4\pi q_m}{\varepsilon} \tag{43}$$

$$\Delta E^c = -\frac{8\pi \mu q_m}{\varepsilon} \tag{44}$$

For most organic molecules the functional group giving rise to the permanent dipole moment resides at a distance from the metal surface where the field, due mainly to the charge distribution in the diffuse layer, is small and the permittivity is relatively high. Hence the effect of potential and charge on their free energy of adsorption is small, and a symmetrical θ–V curve results. The existence of π-bond interaction with the surface does not alter this situation substantially unless the energy of interaction is appreciably field dependent.

The results obtained for the adsorption of n-decylamine and naphthalene on a number of metals are shown in Table 1 below to illustrate the above arguments. Several points become apparent upon examination of these results.

1) The potential of maximum adsorption E_m is in all cases studied here cathodic to the potential of zero charge $E_{q=0}$, as observed also in the case of mercury.[170] This is in accordance with the theory of Bockris, Devanathan and Muller[112] on the structure of the double layer and indicates stronger interaction between metal and the water molecules when the oxygen is toward the metal.

2) The difference $E_m - E_{q=0}$ is not the same for all metals, indicating that ΔE^c may depend on the metal. Bockris and Swinkels[140] have been able to obtain reasonable values for ΔE^c, in agreement with experiment, on the basis of image force and dispersion interactions. They were unable, however, to account quantitatively for the variation of ΔE^c with metal. This may be ascribed to the uncertainty in the values of both $E_{q=0}$ and E_m on solid metals rather than to a weakness of the theory.

3) Most striking perhaps in this table is the fact that the standard free energy of adsorption on five metals and for two very

Table 1

Quantities Relevant for the Adsorption of n-Decylamine and Naphthalene[140,141]

Metal	p.z.c., volt	$E_{max}(\theta \to 0)$, volt		ΔG^0_{ad} at $E_m(\theta \to 0)$, kcal/mole	
		n-decylamine	naphthalene	n-decylamine	naphthalene
Ni	−0.47	−0.7	−0.8	−6.8	−6.0
Fe	−0.50	−0.7	−0.7	−6.6	−7.0
Cu	−0.20	−0.9	−0.9	−7.3	−7.0
Pt (pH = 2)	+0.50*	—	+0.1	−7.4	−8.4
Pt (pH = 12)	−0.04*	—	—	−7.4	−8.4
Pb	−0.70	−1.3	—	−6.2	—

*The data for the p.z.c. on Pt are taken from recent studies in this laboratory.[94,172]

different compounds is almost the same. It is very strongly implied by these results that physical adsorption occurs in all these cases. The fact that naphthalene, which replaces six water molecules on the surface, has a value of ΔG_A^0 similar to that of n-decylamine (assumed to replace only one water molecule)[140] implies that this quantity ($-\Delta G_A^0 = 6$–8 kcal/mole) is characteristic of the replacement on the surface of a water molecule by one or two —CH_2 groups, irrespective (to a first approximation) of the overall size of the molecule.

4) The potentials of maximum adsorption given in Table 1 were values extrapolated to zero coverage by organic. The fundamental variable in electrosorption is the charge q_m rather than the potential. Adsorption studies on mercury show that while the potential of maximum adsorption may vary with coverage, the charge at which adsorption reaches a maximum is independent of it.[170] A substantial dependence of E_m on θ is indicative of a contribution of the dipole potential of the organic to the field in the compact double layer.

5. Adsorption of Organic Compounds on Platinum Electrodes

(i) Carbon Monoxide

The electrosorption of CO on platinum electrodes on open circuit has been measured by galvanostatic charging curves (cf. Figure 4).[156] The coverage, determined from the charge required to

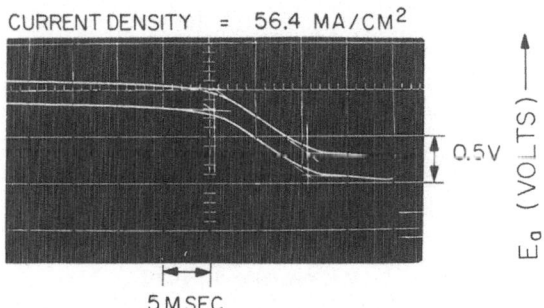

Figure 4. Typical galvanostatic anodic charging curve for CO in $1 M$ H_2SO_4 on Pt black [from Warner and Schuldiner, *J. Electrochem. Soc.* **111** (1964) 992].

oxidize the adsorbed CO compared with that required to deposit a monolayer of oxygen atoms ($453 \, \mu C/cm^2$ and $426 \, \mu C/cm^2$, respectively), is apparently slightly greater than unity. This was interpreted as an indication for the existence of physically adsorbed CO on top of a complete monolayer of chemisorbed CO. Upon removal of CO from the solution, the charge required to remove the adsorbed species decreased to $428 \, \mu C/cm^2$. This corresponds to one molecule of CO per surface Pt atom, assuming a two electron oxidation of CO to CO_2. However, the number of sites available for adsorption on a heterogeneous surface may be different for different species and the slightly higher charge (about 6%) needed to oxidize the CO adsorbed on the surface may not correspond to CO adsorption in excess of a monolayer. Another study of CO adsorption using the galvanostatic technique reported that $390 \, \mu C/cm^2$ (compared with $210 \, \mu C/cm^2$ required to deposit a monolayer of

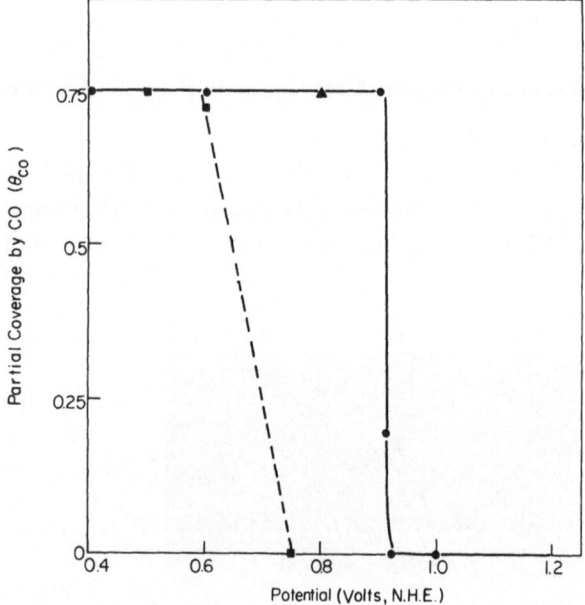

Figure 5. Fraction of Pt surface covered with CO during periodic triangular sweep (0.04 V/sec) as a function of potential: (●) ascending sweep, (■) descending sweep from 1.8 V, (▲) descending sweep from 1.2 V [from Gilman, *J. Phys. Chem.* **66** (1962) 2657].

hydrogen atoms) was required to completely remove the adsorbed CO.[173]

A different approach was to adsorb CO at a constant potential and bulk concentration, then analyze the adsorbed CO with a linear potential sweep.[174] This method indicated a constant coverage of CO of approximately $\theta = 0.75$ between 0.4 and 0.9 V, dropping off rapidly to zero at about 0.91 V (cf. Figure 5). Further results obtained by this method indicated that adsorbed CO has two structures, linear or one-site (two electrons per site) and bridged or two-site (one electron per site) attachment. The contributions of each of these modes of adsorption can be determined from the number of sites occupied by CO (which is proportional to $Q_H^S - Q_H$) and the total charge Q_{CO}^t associated with the adsorbed layer. Thus Gilman obtained the relationship

$$Q_{CO}^t = Q_{CO}^B + Q_{CO}^L = 2(Q_H^S - Q_H) - Q_{CO}^B \qquad (45)$$

where Q_H^S is the number of coulombs/cm² to deposit a monolayer of hydrogen atoms, Q_H is the number of coulombs/cm² obtained in hydrogen deposition in the presence of CO, Q_{CO}^t is the total amount of charge associated with adsorbed CO, Q_{CO}^B is the charge associated with the CO adsorbed by two-site attachment, and Q_{CO}^L is the charge associated with CO adsorbed by one-site attachment. It was concluded that at 30°C, $\theta_{CO}^L = 0$ until θ_{CO}^B values reach the maximum value of 0.44, after which θ_{CO}^L increases to its saturation value of 0.46. Evidence for two types of adsorbed CO has also been obtained from the gas phase adsorption of CO on supported Pt using infrared spectroscopy.

(ii) Formic Acid

The available data on the fractional coverage of platinum anodes by adsorbed formic acid are summarized in Figure 6.[39,176–178] The coverage-potential curves had similar shapes as determined by several investigations; however, the potential-independent portions vary considerably.

From galvanostatic-charging curves with cyclic voltametric measurements, Breiter found that formic acid, adsorbed from 1 M solution, covers a platinum electrode essentially completely ($\theta \approx 1$) between 0.1 and 0.8 V during an anodic potential sweep.[178] However, on the cathodic sweep no formic acid species was

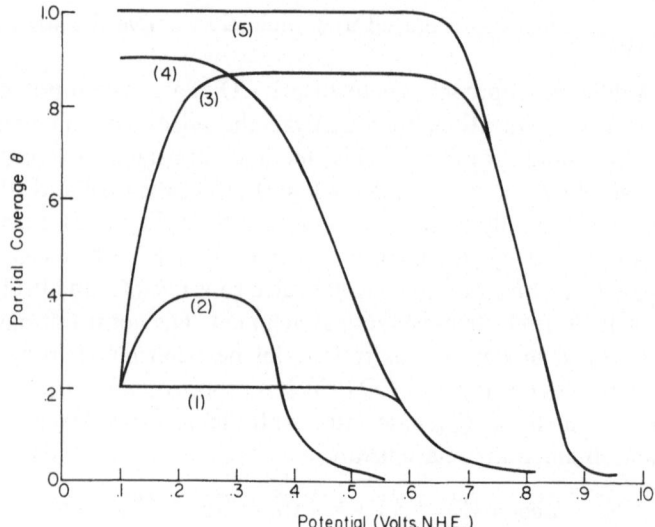

Figure 6. Fraction of Pt surface covered with HCOOH under various experimental conditions: (1) $10^{-4} M$, 30°C, (2) 1 M, 90°C, (3) 1 M, 25°C, (4) 1 M, 40°C, (5) 1 M, 30°C [from Brummer and Makrides, *J. Phys. Chem.* **68** (1964) 1448; Breiter, *Electrochim. Acta* **8** (1963) 447; Fleischmann, Johnson, and Kuhn, *J. Electrochem. Soc.* **111** (1964) 602].

adsorbed until the electrode potential became lower than 0.5 V. Irreversible behavior (with respect to the direction of potential change) was related to the presence of oxygen adsorbed on the electrode which was not removed during the cathodic cycle until a potential of approximately 0.6 V was reached. No coverage by formic-acid species above about 0.9 V was detected and this was also attributed by Breiter to the adsorption of oxygen.[178] The results of Fleischmann *et al.*[177] were similar to those reported by Breiter; however, these authors made corrections for the coverage by hydrogen at potentials more cathodic than about 0.35 V. Brummer and Makrides observed a potential independent adsorption, from 1 M HCOOH solution, only between 0.1 and 0.4 V.[39] Using steady-state measurements, Brummer and Makrides found that adsorption was reversible, i.e., the adsorption recorded as a function of potential in the cathodic direction was identical with that in the anodic direction. From an analysis of the kinetics of adsorption, these authors found that at least 150 sec were required to obtain equilibrium conditions of coverage by formic acid after an anodic pulse (to remove adsorbed formic-acid species) had been applied. (This slow adsorption again points out the disadvantage of non-stationary methods with which equilibrium coverage is usually not attained.) Brummer and Makrides found that their experimental data approximately fit an equation of the form

$$\frac{d\theta}{dt} = k(1 - \theta) - k'\theta \tag{46}$$

They concluded that formic acid is adsorbed according to a Langmuir-type isotherm, which is in agreement with the adsorption of other organic species as will be discussed later.

There has been considerable discussion concerning the nature of the species adsorbed from formic-acid solutions. Several possibilities have been proposed although no definite conclusions have been made, primarily because of a lack of good experimental evidence. This point is further discussed in the section on formic acid oxidation which follows (cf. p. 103).

(iii) Methanol

Pavela was the first to establish direct experimental evidence for the adsorption of methanol on platinized Pt, by removing the electrode from methanol solution and then observing an arrest in the anodic charging curve in 1 N NaOH.[77] He found full coverage for methanol concentrations greater than 0.5 molar, with the coverage depending on concentration according to the Langmuir isotherm. Breiter and Gilman have used a fast potential sweep method and galvanostatic charging to determine methanol coverage on bright platinum anodes in 1 N HClO₄.[102] They observed a linear relation between coverage and the logarithm of the methanol concentration indicating a Temkin-type adsorption isotherm.[179] The coverage was found to be independent of potential from between 0.1 and 0.6 V (at all concentrations of methanol) (cf. Figure 7) and to decrease rapidly at higher potentials. Coverage measurements of surface oxide indicated that the potential at which oxide formation commenced coincided with the potential at which the methanol coverage decreased.[102] This point has been more carefully studied by Bagotzky and Vasilyev who arrived at similar conclusions.[180] These results should, however, be considered very qualitative in view of the inherent limitations and uncertainties

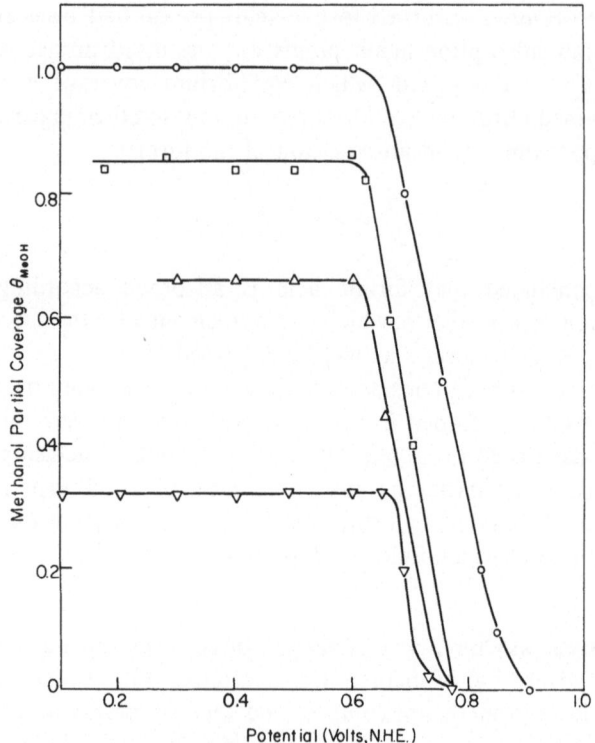

Figure 7. Methanol adsorption as a function of potential during
the anodic sweep (0.03 V/sec): (○) 1 M, (□) 10^{-1} M, (△) 10^{-2} M,
(▽) 10^{-3} M [from Breiter and Gilman, *J. Electrochem. Soc.* **109**
(1962) 626].

involved in the application of the fast potential-sweep technique,
as discussed above [cf. section 3(i)(b)(β)].

(iv) Hydrocarbons

The structure of adsorbed hydrocarbons, and indeed of organic
compounds, has not in general been examined in liquid media.
The adsorption of hydrocarbons on transition metals has been
extensively studied in the gas phase,[181,182] however, although no
general agreement on the structure of adsorbed complexes has been
reached. For example, consider the adsorption of ethylene which
has been studied on nickel, palladium, and supported metal cata-

lysts. The first such study was that of Twigg and Rideal,[183] whose results on the hydrogenation of ethylene at nickel surfaces were best explained if a carbon–carbon double bond broke in chemisorption, giving a complex bound by two-point attachment

$$
\begin{array}{cc}
H \quad\quad H \\
\diagdown \quad\diagup \\
C{=}C \quad + 2Ni \rightarrow CH_2{-}CH_2 \\
\diagup \quad\diagdown \quad\quad\quad\quad | \quad\quad | \\
H \quad\quad H \quad\quad\quad Ni \quad Ni
\end{array}
\qquad (47)
$$

Subsequent work by Conn and Twigg[184] and by Beeck, Smith, and Wheeler[185] confirmed this conclusion. The plausibility of this mechanism depends on the distance between the metal atoms. Twigg and Rideal[186] were able to show that favorable spacings were available on nickel and all other metals active in catalyzing ethylene hydrogenation. Later work by Beeck,[187] using nickel films, suggested that the main surface chemisorption of ethylene involves breakage of C—H rather than C—C bonds:

$$
\begin{array}{cccc}
H \quad\quad H & & H \quad\quad H \\
\diagdown \quad\diagup & & \diagdown \quad\diagup \\
C{=}C \quad + 4Ni \rightarrow H + & & C{=}C \quad + H \\
\diagup \quad\diagdown & \quad\quad | & \quad\quad | \quad | & \quad | \\
H \quad\quad H & \quad\quad Ni & \quad\quad Ni \; Ni & \quad Ni
\end{array}
\qquad (48)
$$

This conclusion was based on the result that when excess ethylene is admitted to a nickel film at room temperatures, ethane appears in the gas phase. Trapnell[188] concluded that the initial chemisorption on tungsten films is a four-site process and the final adsorption self-hydrogenation is a two-site process. Jenkins and Rideal[189] obtained results from self-hydrogenation of ethylene on nickel which fit the equation

$$
10NiH + 10C_2H_4 \rightarrow 3Ni_2C_2H_4 + Ni_4C_2H_2 + 6C_2H_6 \qquad (49)
$$

From magnetic work on nickel powders, Selwood[190] favored a two-bond chemisorption at room temperature. Assuming that the formation of a nickel–carbon bond affects the magnetization of nickel in the same way as a nickel–hydrogen bond, the chemisorption of ethylene on supported nickel at room temperature results in the nickel gaining on the average slightly more than two electrons per molecule. This implies that most of the ethylene is associatively chemisorbed [cf. equation (47)], but that a moderate fraction is held in a dissociative form [equation (48)] requiring four or more sites. In a study of adsorption of hydrogen and several hydrocarbons on iridium using field emission microscopy, Arthur and Hanson[191] found that below 77°K, H_2, C_2H_2, and C_2H_4 are chemisorbed and both C_2H_2 and C_2H_4 are adsorbed in completely saturated form, i.e., no loss of hydrogen. Above 200°K, C_2H_4 decomposed on Ir to C_2H_2 and 2H with the H being desorbed in the temperature range 250–400°K. Bond[182] has concluded that self hydrogenation is absent from ethylene chemisorption at $-78°C$ on Ni films and is of little importance over palladium at this temperature, where most of the ethylene is held by associative attachment, and where the π-bond is broken and two carbon–metal bonds are formed. There is considerable evidence for the associative form existing after ethylene chemisorption on hydrogen-covered nickel–silica and on films when the resulting complex is exposed to hydrogen.[182] There are no corresponding studies on other olefins.

The two modes of ethylene adsorption, i.e., associative and dissociative, have been examined from the point of view of the energetics of adsorption.[32,94] In the gas phase associative adsorption is favored by approximately 25 kcal/mole. In the case of electrosorption two cases may be considered. At low anodic potential (near the reversible potential for hydrogen in the same solution) the hydrogen atoms will remain adsorbed on the surface, while at higher potential they will be ionized. In the former case associative adsorption is favored by about 70 kcal/mole, and in the latter, two modes of adsorption give rise to the same overall heat of adsorption. Such calculations are, however, somewhat limited since the bond energies in molecules adsorbed on the surface may not be the same as in the gas phase. Further support for associative adsorption of ethylene is obtained from a comparison of the kinetics of ethylene electro-oxidation[32] to acetylene electro-oxidation.[193] The RT/F Tafel slope obtained in acetylene oxidation as compared with a $2RT/F$ Tafel slope for ethylene oxidation indicates that the two reactions have different rate-determining steps. This shows that the reacting adsorbed organic complexes are different, otherwise the same rate-determining step would result.

While the structure of chemisorbed acetylene has not been definitely resolved, several studies determining the infrared spectra of acetylene chemisorbed on several metals have been carried out, and were recently reviewed by Eischens.[194] The spectrum observed after chemisorption of acetylene on a palladium-on-silica catalyst showed no evidence for carbon–hydrogen bonds of saturated hydrocarbons. Little, Sheppard, and Yates found absorption bands at 3090 and 3030 cm^{-1} and attributed both to carbon–hydrogen stretching vibrations of olefinic species.[195] The 3090 cm^{-1} band was ascribed to the structure

$$
\begin{array}{c}
H \quad\; H \\
\backslash \quad / \\
C \\
\| \\
C \\
/ \quad \backslash \\
Pd \qquad Pd
\end{array}
$$

and the 3030 cm^{-1} band to

$$
\begin{array}{c}
H \qquad\quad H \\
\backslash \qquad\; / \\
C\!=\!C \\
|\quad\; | \\
Pd\;\; Pd
\end{array}
$$

the latter being the predominant form. Eischens indicates that the latter structure is that which would be predicted for chemisorbed acetylene.

The degree of surface coverage of platinum anodes by hydrocarbons adsorbed from an aqueous solution has recently been examined by several methods. Dahms, Green, and Weber first reported coverage measurements for ethylene adsorbed on platinized gold foil electrodes from 1 N NaOH at 60°C using a radiotracer method.[136] Their results indicated low coverage by ethylene, on the order of $\theta = 0.1$; however, the accuracy of this determination was limited. Griffith and Rhodes used the potential-sweep method and found a surface coverage of platinum by ethylene, adsorbed from 1 N H$_2$SO$_4$ saturated with C$_2$H$_4$ at 80°C, of 0.16 on smooth platinum assuming two-site attachment.[64] The value of this result is subject to the uncertainties imposed by the use of the potential-sweep method as discussed in section $3(i)(b)(\beta)$ (cf. p. 60).

Niedrach has used constant current transients to study the adsorption of several hydrocarbons from acid and alkaline electrolytes at 25°C on supported platinum black electrodes.[196] An

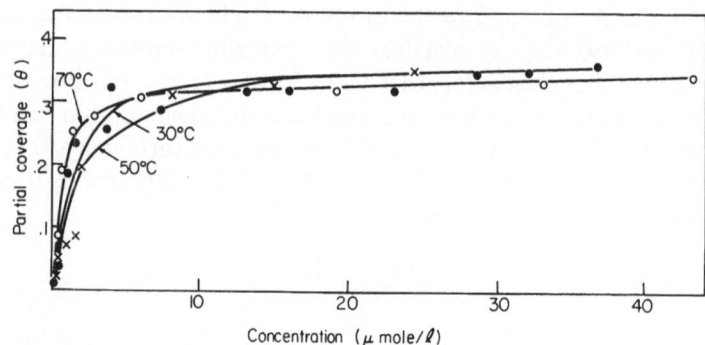

Figure 8. Adsorption isotherms for ethylene on Pt electrodes [from Gileadi, Rubin and Bockris, *J. Phys. Chem.* **69** (1965) 3335].

ethylene coverage of $\theta \approx 0.75$ was found. The volumetric technique, measuring the volume of ethylene adsorbed on equilibration with platinum, has also been used by Niedrach.[196] At the rest potential of ethylene in $5 N\ H_2SO_4$ at 80°C, a coverage of $\theta = 0.77$ was obtained, consistent with the galvanostatic transient measurements. Niedrach observed that no wave attributable to hydrogen appeared in the oxidation curve for ethylene, which indicates little or no dissociative adsorption. On the other hand, the volumetric measurements showed that some of the ethylene adsorbed was hydrogenated on the surface and could be desorbed as ethane.

A recent study[94] of the electrosorption of ethylene on Pt electrodes from 1 N sulfuric acid showed that a limiting coverage of $\theta \doteq 0.4$ was reached when the concentration of ethylene in solution exceeded 10^{-5} moles/liter. The adsorption isotherm obtained at 30°C and a potential of 0.4 V (N.H.E.) is shown in Figure 8. The variation of partial coverage at extreme values of the potential was interpreted in terms of a "competition with water" model proposed by Bockris *et al.*[112,140,141] The low value of limiting coverage is probably due to unavailability of some of the sites for adsorption of a large organic molecule.[94] This may arise because of the geometrical configuration of the surface (small holes or crevices into which the ethylene molecule cannot penetrate) or because of the lack of activity of certain sites on the surface for ethylene chemisorption. In addition, for a molecule taking up n sites on the surface there may be aggregates of $(n - 1), (n - 2) \ldots 1$

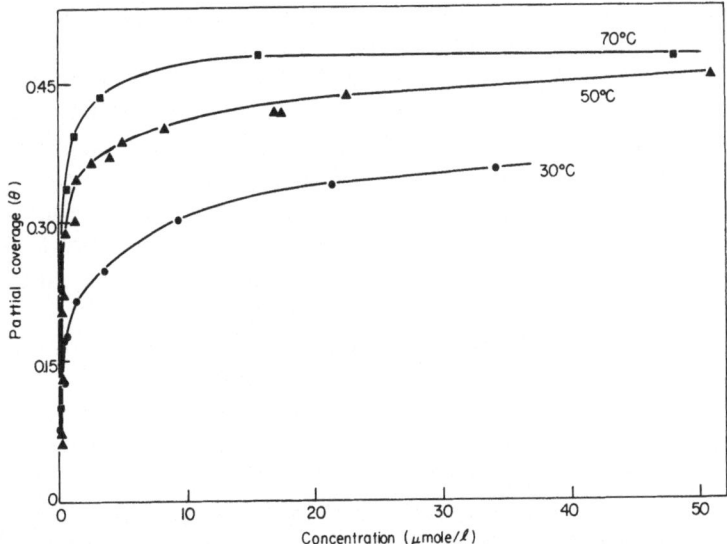

Figure 9. Adsorption isotherms for benzene on Pt electrodes [from Hieland,
Gileadi, and Bockris, *J. Phys. Chem.* **70** (1966) 1207].

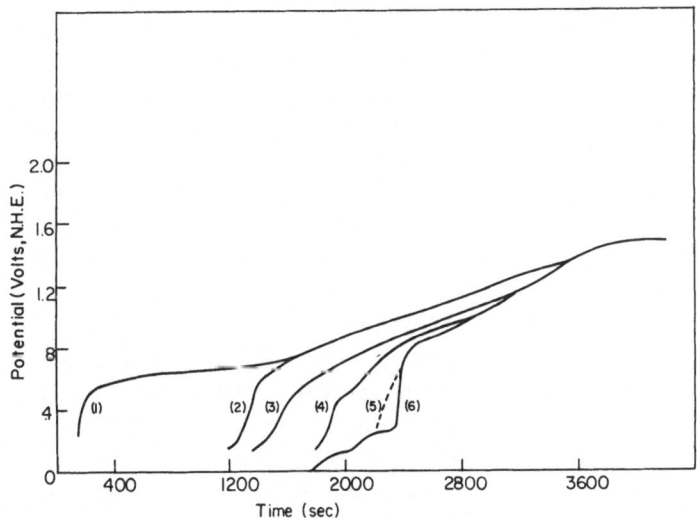

Figure 10. Galvanostatic oxidation curves for hydrocarbons adsorbed on
platinum black electrodes at 25°C from 5 *N* H_2SO_4: (1) C_2H_4, (2) C_4H_{10},
(3) C_3H_8, (4) C_2H_6, (5) CH_4, (6) H_2 [from Niedrach, *J. Electrochem. Soc.* **111**
(1964) 1309].

free sites unavailable for adsorption at any time. Similar behavior of low limiting coverage has been reported previously by Wroblowa and Green[197] in the adsorption of thiourea on gold electrodes. Whatever the cause for the low values of the limiting coverage, one is justified in defining $\theta_{max} = 1.00$ as long as it is established experimentally (cf. Figure 8) that a limiting saturation coverage is reached. Physically this means that the conclusions reached regarding the adsorption behavior apply only to that part of the surface upon which adsorption occurs during the time and in the concentration range studied.

Hieland, Gileadi, and Bockris have recently studied the adsorption of benzene on platinum, using the radiotracer method.[198] Typical results are shown in Figure 9. The coverages indicated here are higher than those obtained with ethylene for a given bulk concentration of the organic.

The adsorption of saturated hydrocarbons has also recently been studied by several methods. Niedrach examined the adsorption of methane, ethane, and propane on platinum using galvanostatic charging curves and also a volumetric method.[196] A comparison of these results is shown in Figures 10 and 11. Niedrach concluded that hydrocarbons fall into several groups when considered in terms of surface coverage. Methane adsorbed to only a small extent, i.e., $\theta < 0.1$. Higher molecular weight saturated hydrocarbons adsorb to a greater extent, e.g., ethane and propane give $\theta \approx 0.2$–0.3. Unsaturated hydrocarbons gave much higher coverages. The electrolyte from which adsorption was carried out had a marked effect on the adsorption behavior of saturated hydrocarbons; however, no detectable effects were observed for unsaturated compounds.

The adsorption of propane onto platinum from concentrated phosphoric acid using anodic galvanostatic charging and cathodic hydrogen deposition has been reported.[199] A typical result is given in Figure 12. It was concluded that, at least initially, propane requires three sites for adsorption and that equilibrium adsorption is obtained after about 1 min, with a maximum coverage between 0.3–0.5.

The adsorption of butane from $1 N H_2SO_4$ on platinum films sputtered onto mica windows has been carried out using the radiotracer technique.[200] Adsorption maxima were observed at

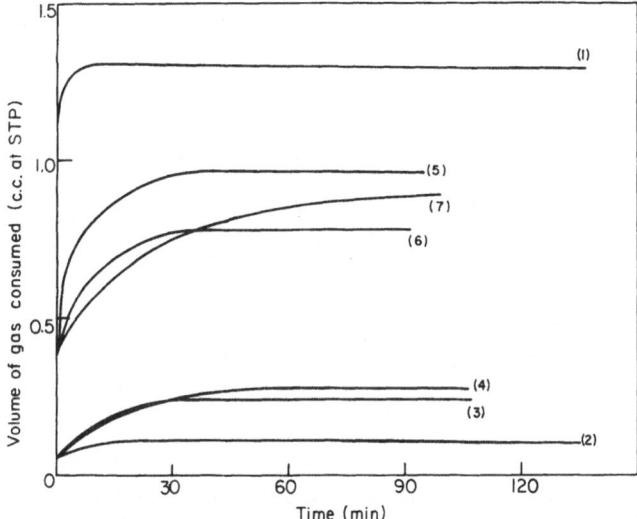

Figure 11. Gases consumed during equilibration with platinum black electrodes: (1) hydrogen, (2) methane, (3) ethane, (4) propane, (5) ethylene, (6) propylene, (7) cyclopropane [from Niedrach, *J. Electrochem. Soc.* **111** (1964) 1309].

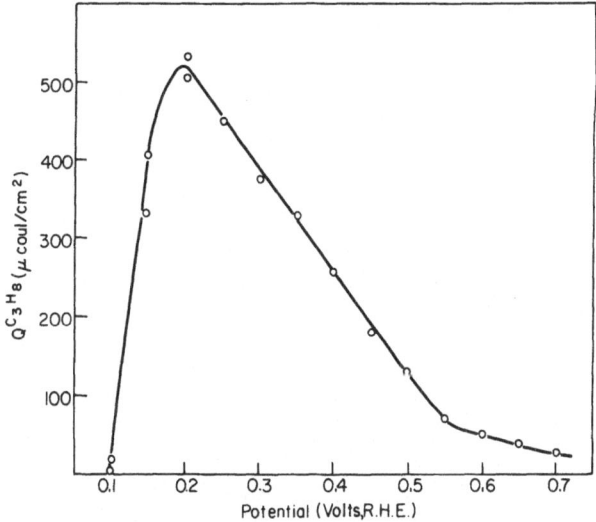

Figure 12. Surface coverage by propane in terms of $\mu C/cm^2$, adsorbed on platinum from 80% H_3PO_4 at 80°C [from Makrides, Brummer, and Dienst, Tyco Labs. Inc., Waltham, Mass., *Second Interim Technical Report*, May, 1964].

0.3 V (N.H.E.) for both 5×10^{-4} and 5×10^{-5} molar radio-n-butane, with the count rate at the adsorption maximum of the former concentration approximately twice that of the latter (Figure 13). A plot of the count rate *vs.* $\sqrt{t}$ gave linear behavior for the first 15–20 min. [This is comparable with the results obtained by Gileali, Rubin, and Bockris for ethylene adsorption (Figure 14)], indicating diffusion controlled adsorption.[94] Ethane was also examined in the same study[200] by galvanostatic charging, and the maximum coverage was indicated to be less than $\theta \sim 0.25$, with no adsorption below 0.225 V.

The isotherms determined experimentally by radiotracer measurements for the adsorption of ethylene on platinum are in agreement with Langmuir-type adsorption.[94]

The form of the Langmuir isotherm may depend on the number of sites involved for ethylene adsorption. From steric considerations, it is probable that four platinum atoms are required for the

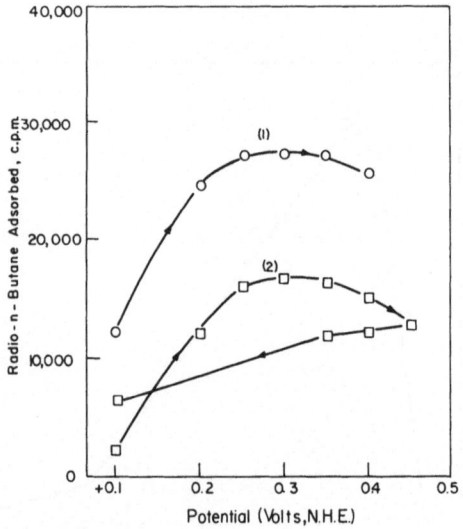

Figure 13. Electrosorption of radio n-butane from $1 N$ H_2SO_4 onto platinum films: (1) $5 \times 10^{-4} M$, (2) $5 \times 10^{-5} M$ [from Flannery, Aronowitz, and Walker, American Oil Company, Research and Development Dept., Whiting, Ind., *Progress Report*, No. 1, March 1965].

adsorption of one molecule of ethylene. Further evidence for four-site attachment is obtained from considerations of covalent bonding with the d-electron vacancy of platinum. Platinum has approximately 0.45 d-electron vacancy per atom, indicating that on the average, two platinum atoms are required to accept one electron from the ethylene. For breaking of the ethylene double bond, two electrons are freed for covalent bonding with platinum, thus requiring four platinum atoms. The role of d-electrons and d-electron vacancies for bonding in transition metals has been discussed by Bond.[182] The role of unpaired d-electrons for oxygen adsorption on oxide-free electrodes has been examined recently by Rao, Damjanovic, and Bockris.[201] The maximum coverage of the several metals studied appeared to be proportional to the number of unpaired d-electrons per atom, to the limit of a full monolayer. These results strongly suggest that the unpaired d-electrons directly participate in the bonding of oxygen atoms adsorbed on noble

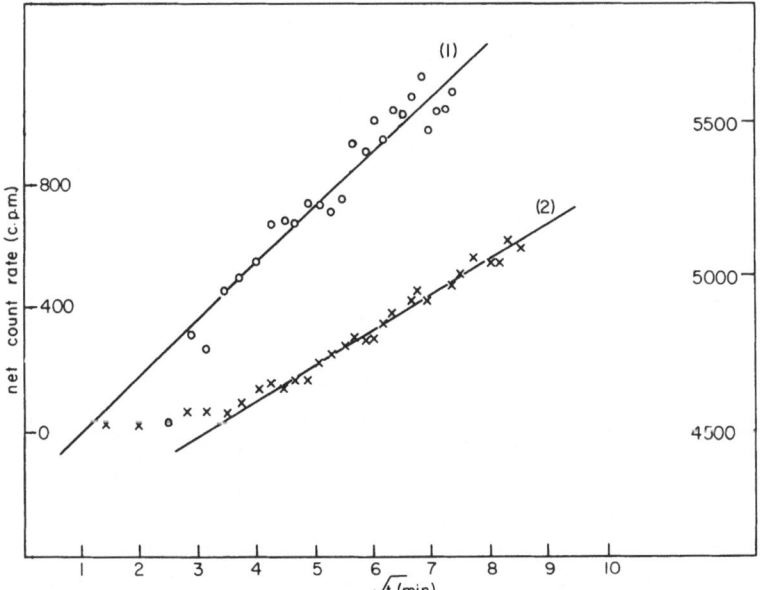

Figure 14. The rate of electrosorption of ethylene onto platinum from $1\,N$ H_2SO_4 at $0.4\,V$ (N.H.E.): (1) $3 \times 10^{-6}\,M$, (2) $3 \times 10^{-7}\,M$ [from Gileadi, Rubin, and Bockris, *J. Phys. Chem.* **69** (1965) 3335].

metals. Thus, in the case of platinum, an oxygen atom shares two electrons with approximately four platinum atoms. For gold, which has no unpaired d-electrons, essentially no adsorption was obtained.

The electrosorption of saturated compounds is usually accompanied with cleavage of C—H and C—C bonds and possibly partial oxidation. Thus, in the study of butanol adsorption on Pt[200] irreversible behavior was observed upon changing the potential between 0.1 and 0.3 V (R.H.E.). Initially, a substantial increase of adsorption was observed in going from 0.1 to 0.3 V, but the lower value could not be reproduced when the potential was shifted back to 0.1 V. After several cycles, the coverage settled at a value intermediate between the initial values for 0.1 V and 0.3 V and became independent of potential. This behavior was interpreted as being due to a rather refractory carbonaceous residue which was formed during slow oxidation at the higher potential. Evidence for the cracking of saturated hydrocarbons during electrosorption was reported by several other authors.[196,200] This is, in fact, to be expected since there is no other way in which a saturated hydrocarbon could be chemisorbed on the surface. The accumulation on the surface of a partially dehydrogenated hydrocarbon residue may give rise to some of the decay of activity observed in anodic oxidation. The reactivation of the surface caused by raising the potential to values near (but below) oxide formation has been attributed to the cleaning of the surface from such refractory residues which have accumulated at lower potentials.[158,202]

IV. MECHANISM OF ORGANIC OXIDATION

1. Carbon Monoxide

Although carbon monoxide is the simplest organic compound which can be anodically oxidized, possessing the advantages of structural simplicity and chemical stability, it has been little studied. Gilman first reported the anodic oxidation of CO in 1962, using cyclic voltametric techniques.[174] His initial observations were extended and further discussed.[203] Another investigation, using galvanostatic charging methods, has been reported by Warner and Schuldiner.[156]

(i) Overall Reaction

The overall reaction for the anodic oxidation of carbon monoxide to CO_2 is

$$CO + H_2O \rightarrow CO_2 + 2H^+ + 2e \qquad (50)$$

The standard reversible potential calculated for this reaction is approximately $+0.10$ V (N.H.E.).

(ii) Open-Circuit Potentials

The rest potential observed on smooth Pt in $1\,N\,H_2SO_4$ at 25°C for CO was 0.35 V (N.H.E.) and independent of CO partial pressure from 0.0037 to 1 atm.[156] The value in the absence of CO, i.e., with helium, was 0.29 V (N.H.E.). In $1\,N\,HClO_4$ at 30°C, the rest potential of CO on smooth Pt was reported to drift slowly from 0.4 to 0.1 V (N.H.E.).[174] This drift was attributed to hydrogen on the electrode which was detected if open-circuit conditions were maintained for periods greater than 20 min.

(iii) Current–Potential Relations

Current–potential curves have been determined only by the cyclic voltametric method using a sweep rate of 40 mV/sec (cf. Figure 15).[203] In the anodic sweep, practically no current flows from 0.4 to 0.9 V. At 0.91 V a vertical rise of the current is observed, and from 0.91 V to 1.6 V the current decreases linearly with potential. The current observed at 0.91 V was dependent on stirring rate

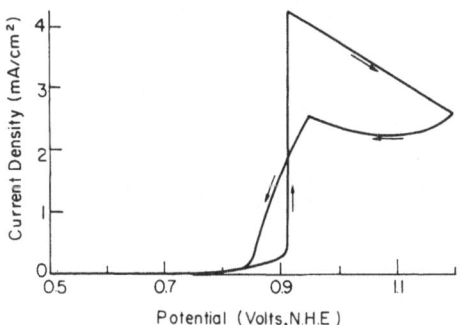

Figure 15. Anodic oxidation of CO using a triangular potential sweep [from Gilman, *J. Phys. Chem.* **66** (1962) 2657].

and was indicated to be controlled by diffusion of CO to the electrode. It was suggested that high coverage by CO prevents any significant oxidation below 0.91 V. Above this potential the coverage decreases rapidly thus permitting the rate of oxidation to increase. However, as the coverage by CO falls to zero the rate of adsorption, controlled by diffusion of CO to the anode, becomes rate-limiting.

(iv) Coverage

The coverage of platinum electrodes by carbon monoxide in CO saturated H_2SO_4 solutions appears to be essentially unity in the potential range 0.3–0.9 V (R.H.E.) and drops sharply to a very low value at 0.91 V.

(v) Mechanistic Conclusions

From Figure 4, it is seen that the electrode potential changes relatively little as θ_{CO} is reduced to zero, suggesting that direct electrochemical oxidation of CO does not occur (in that case the potential would be expected to increase more rapidly as θ_{CO} decreased). This leads to the following proposed mechanism:[156]

$$Pt + H_2O \rightarrow PtO + 2H^+ + 2e \qquad (51)$$

$$Pt + CO \rightleftharpoons PtCO \qquad (52)$$

$$PtO + PtCO \rightarrow Pt_2CO_2 \qquad (53)$$

$$Pt_2CO_2 \rightarrow 2Pt + CO_2 \qquad (54)$$

Equation (51) represents the slow step, otherwise coverage by oxygen would increase with time. (The flatness of the region of CO oxidation on the charging curve at all charging rates indicates that the amount of adsorbed oxygen remains effectively constant until all the CO is removed.) The authors give no further discussion of this mechanism. Equation (51) should be further divided into two steps:

$$Pt + H_2O \rightarrow PtOH + H^+ + e \qquad (55)$$

followed by

$$PtOH \rightarrow PtO + H^+ + e \qquad (56)$$

or

$$2PtOH \rightarrow Pt + PtO + H_2O \qquad (57)$$

PtO is probably the oxidizing species in the potential region considered (0.9–1.2 V) (the presence of adsorbed CO could increase the potential at which PtO is formed). There is no definite evidence to establish the structure of the oxygen species in different potential regions; however, the presence on Pt of several forms of surface oxide has been reported by several investigators.[100,180,204,205] Will and Knorr found three types of adsorbed oxygen corresponding to the potential regions 0.8–1.2 V, 1.2–1.6 V, and > 1.6 V.[100] Bagotzky found three arrests in the oxygen range of the anodic charging current in 1 N H_2SO_4 at 0.6–0.8 V, 0.9–1.1 V, and 1.3–1.6 V.[180] From a study of the mechanism and kinetics of platinum oxide film formation, Feldberg et al. have proposed a sequence similar to equations (55) and (56), with the first electron transfer being the slow step and (56) being a fast reversible reaction.[205] Thus water discharge giving adsorbed hydroxyl radicals is probably

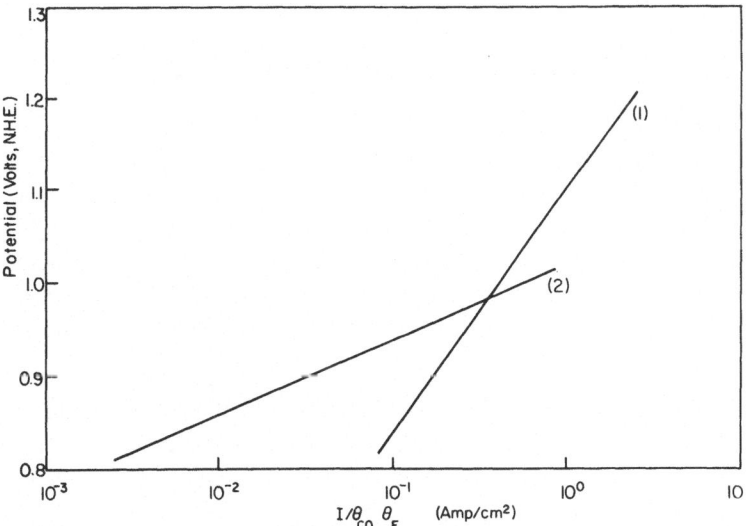

Figure 16. Potential dependence in the anodic oxidation of CO on platinum: (1) 2-site adsorption, (2) 1-site adsorption of CO [from Gilman, *J. Phys. Chem.* **68** (1964) 70].

the rate-limiting step for this proposed mechanism and could be the reacting species rather than PtO.

From a consideration of various models for surface coverage in terms of CO "ad-sites" (θ_{CO}) and "clusters of free sites" (θ_F), Gilman obtained an empirical relation which could be fitted to his experimental data.[203] Plots of $\log(I/\theta_{CO}\theta_F)$ vs. potential gave straight lines for the potential region 0.85–1.1 V (cf. Figure 16), with slopes of about 80 mV for the assigned coverage by the singly bound CO and 130 mV for the CO adsorbed at two sites per molecule. These slopes do not, however, permit a clear interpretation as diagnostic criteria. From his analysis, Gilman has proposed a "reactant-pair" mechanism which postulates the reaction of adsorbed CO with H_2O adsorbed at an adjacent site. The adsorption of CO was considered rapid and the first electron transfers from the adsorbed CO—H_2O complex was supposed rate-determining. On the basis of available experimental evidence, it is not possible to distinguish between Gilman's "reactant-pair" mechanism and the water-discharge mechanism proposed by Warner and Schuldiner; however, those mechanisms are essentially similar in nature.

2. Carboxylic Acids, Low Potential Region

(i) Formic Acid

Müller made the first notable mechanistic investigation of the anodic oxidation of formic acid in 1923, obtaining current–potential curves on several platinum metals over a wide pH range.[23,24,26] Most of the recent electro-organic investigations have considered formic acid because it represents one of the simplest reactions, requiring only two electron transfers per molecule for complete oxidation.

(a) Overall reaction—The overall reaction for the anodic oxidation of formic acid in acidic solutions has been established by a number of investigators, beginning with the work of Müller,[23,24,26] as the following:

$$HCOOH \rightarrow CO_2 + 2H^+ + 2e \qquad (58)$$

with virtually 100% conversion to CO_2. There is some controversy in the literature as to the reaction in basic media. A few investigators indicate uniform reaction throughout the pH range,[23,24,26,39] while others have indicated that formic acid (or the formate ion) is not oxidized in alkaline solutions.[52,53,77,206,207] A maximum rate of oxidation has been reported at pH values near the pK of formic acid.[208,209] An increase of rate with pH up to a pH of 4 to 5 (pK of HCOOH is 3.9), after which the rate remained constant with further increasing pH, has also been reported.[180] As a result of these experimental differences, definite conclusions cannot be made concerning the reaction of formic acid in basic solutions. Further, since most of the mechanistic investigations have been limited to acidic solutions, this discussion will necessarily be confined to the acidic region.

(b) *Open-circuit potentials*—The standard reversible potential calculated from thermodynamic data for formic acid oxidation [equation (58)] is $E^0 = -0.196$ V (N.H.E.). Rest potentials in the range of 0.08–0.17 V (N.H.E.) have been observed at pH ≈ 1.[210,211] These results were obtained by anodically polarizing the electrode, allowing the potential to decay, and observing the potential minimum. The wide range of rest potentials observed for a given solution are explained by the pretreatment of the electrode (quantity of oxide formed, etc.). The potential was found to increase from the minimum by 0.1–0.2 V, with time, and the final constant value reached depended on experimental conditions. Stirring of the solution with nitrogen or carbon dioxide increased the rate of potential increase from the minimum, but approximately the same final values were obtained. Stirring with CO gave a pronounced increase in rate and also established higher constant potentials than either CO_2 or N_2. The rate of increase from the minimum potential was also found to be lowered with increasing pH. Various parameters determined for the rest potential behavior are given in Table 2.

The minimum in the open-circuit potential has been attributed to a mixed potential between hydrogen, resulting from dissociation of formic acid, and some organic species.[211] Gottlieb[210] found that dissociation of formic acid was not a significant factor in the behavior of the system and favored the overall scheme:

$$HCOOH \rightarrow CO_2 + 2H^+ + 2e \qquad (58)$$

$$2H^+ + 2e \to H_2 \tag{59}$$

He observed that the rate of CO_2 evolution decreased markedly at potentials more negative than the open-circuit potential, indicating that CO_2 formation at open circuit is an electrochemical process and that the rest potential is a mixed potential. In their study of the open-circuit potential decay reaction of preanodized platinum black electrodes with formic acid in acid solutions, Oxley et al.[212] observed a slow potential decay to 0.8 V (N.H.E.) characterized by a delay time τ. At 0.8 V, the potential fell rapidly to values between 0.05 and 0.2 V, depending on the experimental conditions. τ was independent of stirring in the concentration range 5×10^{-3} to $1\,M$ HCOOH, but was inversely proportional to the bulk concentration, i.e., $\tau = K/C_{HCOOH}$. τ decreased with increasing temperature in the region 30°–60°C. The quantity of CO_2 evolved in open-circuit decay was determined with a special manometer system and found to be coulombically equivalent to the oxygen coverage. These observations led to the following scheme for the open-circuit decay:

$$PtO + HCOOH \to CO_2 + H_2O + Pt \tag{60}$$

$$PtO + 2H^+ + 2e \to Pt + H_2O \tag{61}$$

$$HCOOH \to CO_2 + 2H^+ + 2e \tag{58}$$

It should be noted here that equations (58) and (61) do not imply two-electron transfer steps but represent overall reactions which involve two or more steps. The initial decay giving rise to the transition time τ is supposed due to a chemical reduction of the

Table 2
Parameters for Rest Potential in Solutions of Formic Acid

$dE/d\log[\text{HCOOH}]$	$dE/d[\text{CO}_3^-]$	$dE/d[\text{CO}_2]$	$dE/d\,\text{pH}$	Ref.
0.032	0	0	$-RT/F\ (0 \leqslant \text{pH} \leqslant 6)$ (deviation from this at pH > 6)	210
—	—	0	$-RT/F\ (2 \leqslant \text{pH} \leqslant 7)$ (deviation for pH < 2, pH < 7)	211

oxide by formic acid. At lower potentials cathodic reduction of the oxide occurs and an accelerating reaction develops which is a mixed process involving equations (58) and (61) until all the oxide is reacted. While this scheme explains the short time behavior, it does not account for the stable rest potentials which could not involve reaction with oxide. In a study of open-circuit potentials, Frumkin[213,214] proposed a redox system established from adsorbed substances and components of the solution. He argued that since the exchange current of the organic species is low, the equilibrium potential should be essentially controlled by either

$$H_{ad} \rightleftharpoons H^+ + e \qquad (62)$$

or

$$H_{ad} + OH^- \rightleftharpoons H_2O + e \qquad (63)$$

the H_{ad} resulting from dissociative adsorption of the organic. Breiter and Gilman had previously arrived at similar conclusions in their study of methanol;[102] however, other studies have indicated that hydrogen does not result from dissociation of the formic acid molecule.[210] In view of the contradictory observations reported, the nature of the open-circuit potential is still open to question. It appears that the rest potential may be established by an oxidation-reduction equilibrium which probably involves some type of hydrogen species.

The increase of the open-circuit potential from the minimum has been attributed to adsorption of an organic species generally assumed to be CO. This was concluded primarily from the observed effect of CO on the establishment of the constant rest potential.[211] Other investigators, particularly Giner,[215] have favored CO as the poisoning species. The evidence for this proposal is not conclusive, however, and further experimental investigation is necessary to establish the mechanism of the open-circuit potentials. (The fact that a comparison of cyclic voltametric current–potential curves for formic acid and "reduced CO_2," which may be assumed to be CO, shows similarity of the shape of the adsorption peaks, sweep-rate dependence, and temperature dependence is not conclusive evidence since most adsorbed organic species which are oxidized in this potential region exhibit similar curves.) As shown by Schuldiner and Warner,[109] a further cause of this behavior could be the presence of hydrogen in the metal resulting from cathodic pretreatment of electrodes. In this case the hydrogen initially present would give a lower open-circuit potential but could be rapidly removed with the eventual establishment of the open-circuit potential characteristic of the organic.

(c) *Current–potential relations*—Steady-state potentiostatic[39,177] and galvanostatic[210] current–potential curves exhibit a Tafel region in the potential range 0.15–0.45 V (N.H.E.), with a limiting current above 0.45 V until the oxygen region is reached. Tafel slopes of $2RT/F$ at low temperatures (25°C), and RT/F at high temperatures (90°C) are reported with intermediate slopes at intermediate temperatures. Slopes of $2RT/F$ are also reported from cyclic voltametric measurements where this data can be obtained. Polarization curves obtained in cyclic voltametry generally give three oxidation peaks in the anodic cycle for formic acid, which occur typically at 0.4–0.6 V (N.H.E.) with a current maximum at 0.5 V, 0.7–1.0 V with a maximum at 0.9 V, and 1.2–1.6 V with a maximum at 1.4 V (cf. Figure 17).[216]

(d) *Exchange current density*—Values of the exchange current density are given in Table 3, along with other pertinent information. A considerable spread of values from 10^{-7} to 10^{-15} A/cm^2 is observed. There is not sufficient information to establish the proper value (or values, since the change in Tafel slope probably indicates a change in mechanism).

(e) *Other kinetic parameters*—The pH dependence of the rate of formic acid oxidation has been examined by steady-state measurements and also by the shift of the current peak in cyclic voltametric studies. The dependence of rate on formic acid concentration has been determined by similar methods and also by variation of the area under the current peak with concentration in cyclic voltametry. These parameters are all collected in Table 3.

(f) *Evidence for type of species adsorbed*—Differences in experimental observations have led to a controversy on the identity of the species adsorbed in the formic acid oxidation. Table 4 presents a summary of these observations.

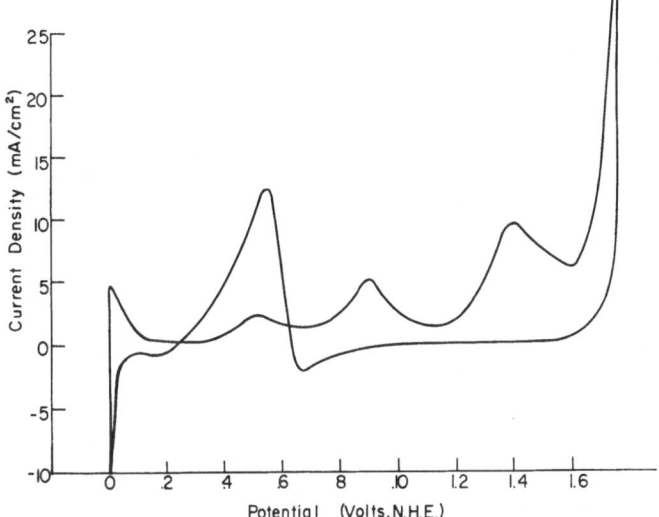

Figure 17. Typical current–potential curve for anodic oxidation of formic acid at platinum with potential sweep rate $dV/dt = 0.275$ V/sec [from Kutscher and Vielstich, *Electrochim. Acta* **8** (1963) 985].

In view of the contradictory observations reported, it is clearly not possible to establish the reactive adsorbed species; however, the formic acid molecule is favored in acidic solutions.

(g) *Mechanistic conclusions*—Although significant differences have been reported in the experimental observations for the anodic oxidation of formic acid, most investigators have preferred the free radical or dissociative mechanism. This mechanism, which is represented by the following scheme, has not, however, been properly justified:

$$HCOOH_{soln} \rightleftharpoons HCOOH_{ad} \qquad (64)$$

$$HCOOH_{ad} \xrightarrow{r.d.s.} HCOO_{ad} + H_{ad} \qquad (65)$$

$$H_{ad} \rightleftharpoons H^+ + e \qquad (62)$$

$$HCOO_{ad} \rightarrow CO_2 + H^+ + e \qquad (66)$$

Following a suggestion by Frumkin,[213,214] Bagotzky and Vasilyev[180] have proposed that a purely chemical surface reaction such as step (65) in the above scheme could give rise to a linear

Table 3

Kinetic Parameters for HCOOH Oxidation

T, °C	b	C_{HCOOH}, mole/liter	pH	i_0, A/cm²	$\left(\dfrac{\partial \log i}{\partial\, pH}\right)_v$	$\left(\dfrac{\partial \log i}{\partial \log C_{HCOOH}}\right)_v$	Ref.
23	$2RT/F$	—	—	—	~0.5	~1.0	207
23	$2RT/F$	1.0	~0.3	2×10^{-10}	—	—	210
25	$2RT/F{-}3RT/F$	0.0057	~0.3	$\sim 10^{-10}$	—	—	217
25	$2RT/F$	—	—	—	~1.0	~0.5	218
25	$2RT/F$	0.55	4.26	$\sim 10^{-9}$	~0.5	~0.5	180
25	$2RT/F$	—	—	—	—	—	177
30	RT/F	3×10^{-3}	~0	4×10^{-7}	—	—	178
30	$RT/3F$	1×10^{-3}	~0	5×10^{-9}	—	—	178
40	$\sim RT/F$	1.0	1.0	2×10^{-15}	~1.0	~1.0	39
90	RT/F	1.0	~0	2×10^{-10}	—	—	177
—	—	—	—	—	—	~1.0	206

Tafel relationship on a heterogeneous surface, due to electro-chemical equilibrium in the following hydrogen adsorption step (62). The argument is that the total surface coverage will depend on potential due to the dependence of the hydrogen coverage θ_H on potential and this will determine the rate of the rate-determining step (65) under Temkin conditions.[90,179,219] It is clear, however, that this could only apply at low values of the potential [$E < 0.3$ V (R.H.E.)] where coverage by hydrogen is appreciable ($0.2 < \theta_H < 0.8$). This is below the potential at which most organic fuels can be readily oxidized anodically. The effect of following steps has been considered for both equilibrium and nonequilibrium steps for the hydrogen evolution reaction[220,221] and for a general reduction reaction involving only one electron transfer.[222] Frumkin has shown, in a recent paper,[223] that the adsorption of organic sub-stances on metals which adsorb hydrogen will be potential depen-dent, assuming dissociative adsorption. This, however, is also applicable only for intermediate values of coverage by hydrogen assuming Temkin-type adsorption.

Consider the scheme given above [equations (64)–(66)] for formic acid oxidation. The rate-determining step (65) may be regarded as an adsorption step, since it results in an increase in the number of sites occupied on the surface. The corresponding rate equation, under Temkin conditions[179,219] will be

$$\vec{v} = \vec{k}\theta_1(1 - \theta_T)\exp\left[\frac{\alpha f(\theta)}{RT}\right]\exp\left[\frac{-2(1 - \alpha)f(\theta)}{RT}\right] \quad (67)$$

where

$$f(\theta) = r_1\theta_1 + r_2\theta_2 + r_3\theta_3 \quad (68)$$

and r_1, r_2, and r_3 are the Temkin parameters for the three kinds of adsorbed species involved in this reaction scheme, HCOOH, HCOO, and H, respectively. The function $f(\theta)$ represents the rate of change of the apparent standard free energy of adsorption ΔG_θ^0 with total coverage θ_T[219]

$$\Delta G_\theta^0 = \Delta G_{\theta=0}^0 - f(\theta) \quad (69)$$

Table 4
Evidence for the Nature of the Adsorbed Species

Species	Observation	Ref.
HCOOH	Optimum acidity for oxidation is in the range of pH 0–1.	207
HCOOH	Addition of HCOOH caused no change in cyclic polaro-grams in $2N$ NaOH while oxidation peaks were observed in $2N$ H_2SO_4.	206
HCOOH	Observed current densities for HCOOH oxidation are much larger than can be accounted for by the calculated limiting diffusion currents of other species [$HCOO^-$, $(HCOOH \cdot HCOO)^-$, $(HCOOH)_2$, etc.].	178
not $HCOO^-$	In anodic oxidation, methanol is converted almost quantitatively to formate in basic solution but completely to CO_2 in acid solution.	52 53 77 207
not $HCOO^-$	The coverage decreases with increasing anodic potential making the adsorption of a negatively charged species unlikely.	39
$HCOO^-$	Infra-red spectra for HCOOH adsorption in the gas phase on platinum indicate $HCOO^-$ as the species adsorbed.	312
not $HCOO\cdot$	The maximum charge observed for adsorption from anodic charging ($q = 210\,\mu C/cm^2$) would imply an area of formate radical of $7\,Å^2$ as compared with $\sim 14\,Å^2$ for HCOOH.	39
CO or $HCOO\cdot$	From catalytic decomposition, $HCOOH \rightarrow CO + H_2O$. Comparison of cyclic polarograms for HCOOH, and "reduced CO_2" indicates similarity of the adsorbed species.	215
not CO	From anodic transients, charge corresponding to full coverage for CO ($\sim 340\,\mu C/cm^2$) is higher than adsorption for HCOOH ($\sim 210\,\mu C/cm^2$), and the anodic traces have significantly different shapes.	39

From equilibrium in the first adsorption step (64) one obtains under Temkin conditions, at intermediate values of the coverage,

$$f(\theta) = RT \ln Kc_1 \qquad (70)$$

Equation (70) reduces to the more familiar form of the Temkin isotherm[179]

$$\theta_1 = \frac{RT}{r_1} \ln Kc_1 \tag{71}$$

under the limiting conditions of $\theta_1 \doteq \theta_T$, and hence $f(\theta) \doteq r_1\theta_1$.
Substituting $f(\theta)$ from equation (70) into equation (67) and
taking $\alpha \doteq 1 - \alpha \doteq 0.5$ one has

$$\vec{v} = \vec{k}\theta_1(1 - \theta_T)c^{-0.5} \doteq \vec{k}c^{-0.5} \tag{72}$$

Thus a negative reaction order is predicted, contrary to experiment[180] (cf. Table 3).

Considering the electrochemical equilibrium in step (62) one
has, according to Frumkin's isotherm[91] as applied to this system,

$$\frac{\theta_3}{1 - \theta_T} \exp\left[\frac{f(\theta)}{RT}\right] = Kc_{H^+} \exp\left(\frac{-VF}{RT}\right) \tag{73}$$

At low anodic potentials where $0.2 < \theta_3 < 0.8$ equation (73) may
be written approximately as

$$f(\theta) = -VF + \ln c_{H^+} + \text{const} \tag{74}$$

Combining equations (67) and (74) yields

$$i = Kc_{H^+} \exp\left(\frac{VF}{2RT}\right) \tag{75}$$

where K is a combination of constants which includes the conversion from units of reaction rate to current density.

Equation (75) leads to the experimentally observed Tafel
slope $b = 2RT/F$ for this reaction. It was derived, however, on the
assumption of intermediate coverage by atomic hydrogen, which is
only valid at low anodic potentials [below approximately 0.3 V
(R.H.E.)]. At higher anodic potentials where $\theta_3 \ll \theta_1 ; \theta_2$ it is clear
that $f(\theta)$ will be essentially independent of θ_3 [equation (68)] and
the overall reaction rate will not depend on potential.

It is more reasonable to assume that surface dissociation of the
adsorbed formic acid molecule and the following ionization of the
hydrogen atom formed on the surface [equations (65) and (62)]
occur as a single step

$$\text{HCOOH}_{ad} \rightarrow \text{HCOO}'_{ad} + \text{H}^+ + e \tag{76}$$

If this step is rate-determining in the reaction sequence, a Tafel slope of $b = 2RT/F$ is predicted, in agreement with experiment. The reaction order will depend on the type of adsorption isotherm involved. Under Langmuir conditions a reaction order of unity is predicted at low coverage decreasing gradually to zero as the coverage is increased. A Freundlich-type isotherm would give rise to fractional reaction order over a range of concentration, while under Temkin conditions the reaction rate will be proportional to $\log c_1$, and will vary little with concentration.

Some investigators have reported a reaction order of unity for this system.[39,206,207] This is in agreement with the above model assuming Langmuir adsorption at low coverage. A reaction order of 0.5 has also been reported[180,218] which would only be consistent with the above mechanism over a narrow concentration range at intermediate values of the coverage under Langmuir conditions, or if the system followed the Freundlich isotherm

$$\theta = Kc^{1/n} \tag{77}$$

with $n = 2$.

Further experiments need be done to clarify this point. One may surmise perhaps that the Langmuir adsorption isotherm applies for electrosorption of formic acid (as is the case of, e.g., ethylene[94]) and the discrepancy between reported values of the reaction order arises due to the gradual change from the linear θ–c relationship observed at low values of the coverage to an approximate Freundlich-like region at intermediate-coverage values.

A mechanism involving water discharge has been suggested by several investigators,[180,213,216] according to the following:

$$H_2O \rightleftharpoons \cdot OH_{ad} + H^+ + e \tag{78}$$

$$HCOOH_{soln} \rightleftharpoons HCOOH_{ad} \tag{64}$$

$$\cdot OH_{ad} + HCOOH_{ad} \xrightarrow{r.d.s.} \cdot COOH_{ad} + H_2O \tag{79}$$

$$\cdot COOH_{ad} \rightarrow CO_2 + H^+ + e \tag{66}$$

For step (79) rate-determining, and under Temkin conditions one can write

$$\vec{v} = \vec{k}\theta_4\theta_1 \exp\left[\frac{\alpha f(\theta)}{RT}\right] \tag{80}$$

From equilibria in equations (64) and (78) one obtains, respectively,

$$f(\theta) = RT \ln Kc_1 \tag{70}$$

and

$$Kc_{H^+} \left(\frac{\theta_4}{1 - \theta_T}\right) \exp\left[\frac{f(\theta)}{RT}\right] = \exp\left(\frac{VF}{RT}\right) \tag{81}$$

where

$$f(\theta) = r_1\theta_1 + r_2\theta_2 + r_4\theta_4 \tag{82}$$

and the subscript 4 refers to adsorbed OH radicals.

At low coverage by OH radicals and intermediate coverage by adsorbed formic acid, combination of equations (80), (81), and (70) gives

$$\vec{v} = Kc_{H^+}^{-1} c_1^{-0.5} \exp\left(\frac{VF}{RT}\right) \tag{83}$$

The derived Tafel slope $b = RT/F$ is in agreement with some reported observations[39,177,178] and so is the pH effect,[39,218] but the negative reaction order is not found experimentally (cf. Table 3).

At higher anodic potentials $[E > 0.9 \text{ V (R.H.E.)}]$ where coverage by OH radicals is intermediate, equation (81) can be written approximately as

$$f(\theta) = VF - RT \ln c_{H^+} + \text{const} \tag{84}$$

Combined with equations (70) and (80) this yields for the rate expression,

$$v = Kc_{H^+}^{-\alpha} \exp\left(\frac{\alpha VF}{RT}\right) = k'c^\alpha \tag{85}$$

Equation (85) predicts a Tafel slope of $b = 2RT/F$ in agreement with most experimental work,[177,180,207,210,217,218] a reaction order of 0.5, and a pH coefficient $(d \ln i/d \text{ pH})_V$ of 0.5 in agreement with most experimental observations (cf. Table 3). This scheme may apply, however, only at high anodic potentials [about 0.9 V (R.H.E.) and above] which are of little interest for fuel oxidation studies.

Other mechanisms have been suggested, for example, the electron-radical mechanism of Shlygin and Bogdanovsky[224,225] which postulates electron transfer from the organic species with formation of a positively charged ion which subsequently loses a proton to a water molecule and combines with a hydroxyl radical. This sequence may be represented by the following:

$$HCOOH \rightarrow [HCOOH]_{ad}^+ + e \qquad (86)$$

$$[HCOOH]_{ad}^+ + H_2O \rightarrow HCOO_{ad}^\cdot + H_3O^+ \qquad (87)$$

$$H_2O \rightarrow \cdot OH_{ad} + H^+ + e \qquad (78)$$

$$HCOO_{ad}^\cdot + \cdot OH_{ad} \rightarrow CO_2 + H_2O \qquad (88)$$

Assuming that step (88) is rate-determining, and following the treatment used previously with θ_{OH} negligible but $\theta_{[HCOOH]^+}$ in the Temkin region, the rate equation is of the form

$$i = kC_{HCOOH}C_H^{-1} \exp\left(\frac{VF}{RT}\right) \qquad (89)$$

where the derived parameters do not correspond to those obtained experimentally.

In summary, it may be stated that only a mechanism of the type proposed by Bagotzky,[180] with the modification that steps (65) and (62) are assumed to occur simultaneously, can account for the kinetic parameters observed in most investigations of formic acid oxidation. Further experimental work is undoubtedly required before the mechanism of this reaction can be understood.

(ii) Oxalic Acid

Most of the investigations of the anodic oxidation of oxalic acid were carried out in a potential region where oxide is present on the electrode; it was generally held that oxidation does not occur below 0.8 V (N.H.E.). In a recent study, Johnson, Wroblowa, and Bockris[226] have obtained electrode kinetic parameters for the low potential region [0.5–0.8 V (N.H.E.)], and this discussion shall essentially follow that investigation.

(a) Overall reaction—It was reported early that the electrochemical oxidation of oxalic acid in $1 N H_2SO_4$ in the potential region

(0.8–1.10 V) gave carbon dioxide quantitatively according to the reaction[227]

$$H_2C_2O_4 \rightarrow 2CO_2 + 2H^+ + 2e \qquad (90)$$

In the higher potential region[228] and at lower potentials (0.5 V),[226] the complete oxidation of oxalic acid to carbon dioxide in acidic solution was later established by other workers. No oxidation of oxalic acid in alkaline solutions below oxide formation was observed.

(b) *Open-circuit potential*—Calculation from thermodynamic data gives a reversible potential of -0.58 V for reaction (90), assuming $1\ M\ H_2C_2O_4$, $1\ N\ H_2SO_4$, and a temperature 80°C. The observed rest potentials behaved similarly to those found with formic acid. When platinum electrodes polarized at 1.3 V were placed on open circuit in the presence of oxalic acid, the potential rapidly decreased, passing through a minimum at about 0.18 V, and then slowly increased to a steady state of 0.25–0.3 V. Similar behavior was observed when the current was interrupted in the Tafel region.

(c) *Current–potential relation*—Current–potential curves determined by steady-state galvanostatic and potentiostatic methods exhibit linear Tafel regions in the potential range 0.5–0.7 V (N.H.E.), with a slope of 70 mV (RT/F) for Pt anodes in acidic solution at 80°C. Above 0.7 V, the slope increases to a limiting current beginning at about 1.3 V. Polarization above 1.3 V results in passivation of the electrode.[158,226] Extrapolation of the linear Tafel line to the calculated reversible potential gives an exchange current density $i_0 \sim 10^{-20}$ A/cm^2.

(d) *Other kinetic parameters*—The order of reaction with respect to the bulk concentration of oxalic acid has been determined from the shift of Tafel lines with concentration to be $(\partial \log i / \partial \log C_{H_2C_2O_4})_V - 0.35$. The quantitative effect of pH could not be directly determined since the dissociation of oxalic acid in less acidic solution resulted in pH changes. From the assumed rate equation $i = kC_{H_2C_2O_4}^{0.35}C_{H^+}^{x}e^{FV/RT}$, points on the Tafel line were calculated to a different (arbitrary) pH using assumed values of x. The i values at a given V were then plotted as a function of $C_{H_2C_2O_4}$ at constant pH. The proper slope was found only for $x = -0.55$; thus, $(d \log i / d\ \mathrm{pH})_V = 0.55$.

(e) *Activation energy*—Apparent activation energies were calculated from the variation of current with temperature at constant potential to be 22 ± 1 kcal in the Tafel region. Below 50°C, the current fell more rapidly than expected and a determination of the Tafel slope at 30°C indicated a slope of $2RT/F$.

(f) *Evidence for reacting species*—Reactivity of oxalic acid in acid and lack of reactivity in base indicates the undissociated molecule as the reacting species. Further evidence is obtained from the fact that consistent reaction orders were obtained only if the molecule was assumed to be the reactant.

(g) *Mechanistic conclusions*—Several possible mechanisms have been suggested for the anodic oxidation of oxalic acid, including purely chemical oxidation of the acid by platinum oxide,[227,229,230] partial oxidation by the oxide with one electron transfer from the acid,[231,232] and both electron transfers from the acid molecule.[226] Johnson, Wroblowa, and Bockris have discussed a number of possible mechanisms with the derived kinetic parameters,[226] and those leading to the proper parameters are presented below for the high temperature conditions ($b = RT/F$). A mechanism involving both charge transfers from the organic species may be written

$$H_2C_2O_4 \rightleftharpoons HC_2O_{4\,ad} + H^+ + e \tag{91}$$

$$HC_2O_{4\,ad} \xrightarrow{\text{r.d.s.}} 2CO_2 + H^+ + e \tag{92}$$

For Temkin conditions, the rate equation for (92) rate-determining is

$$i = k\theta \exp\left[\frac{\alpha f(\theta)}{RT}\right] \exp\left(\frac{\beta V F}{RT}\right) \tag{93}$$

For step (91) in quasi-equilibrium, $\alpha = 0.5$. Neglecting the pre-exponential coverage terms we obtain

$$f(\theta) = RT \ln K + RT \ln C_{ox} - RT \ln C_{H^+} + FV \tag{94}$$

Substituting and neglecting the pre-exponential coverage term one obtains

$$i = k C_{ox}^{\alpha} C_{H^+}^{-\alpha} \exp\left(\frac{VF}{RT}\right) \tag{95}$$

and for $\alpha = 0.5$ the observed Tafel slope, pH dependence, and concentration dependence are obtained.

A similar mechanism may also be written *viz*.

$$H_2C_2O_4 \rightarrow HC_2O_{4\,ad} + H^+ + e \tag{91}$$

$$HC_2O_{4\,ad} \rightarrow HCO_{2\,ad} + CO_2 \tag{96}$$

$$HCO_{2\,ad} \rightarrow CO_2 + H^+ + e \tag{97}$$

For Langmuir conditions and (96) rate-determining, the proper Tafel slope is derived; however, the concentration and pH dependence are unity rather than 0.5. For Temkin conditions and (97) rate-determining, the proper parameters are derived as above.

Mechanisms involving water discharge are the following:

$$H_2O \rightarrow H^+ + OH_{ad} + e \tag{78}$$

$$H_2C_2O_4 + OH_{ad} \rightarrow H_2O + HC_2O_{4\,ad} \tag{98}$$

$$HC_2O_{4\,ad} \rightarrow 2CO_2 + H^+ + e \tag{92}$$

or

$$HC_2O_{4\,ad} \rightarrow HCO_{2\,ad} + CO_2 \tag{96}$$

and

$$HCO_{2\,ad} \rightarrow CO_2 + H^+ + e \tag{97}$$

For Temkin conditions and (92) rate-determining we have

$$i = k\theta_{HC_2H_4} \exp\left(\frac{\beta VF}{RT}\right) \exp\left[\frac{\alpha f(\theta)}{RT}\right] \tag{99}$$

For quasi-equilibrium in (78) and taking $\theta_{HC_2O_4} \approx \theta_T$, we find

$$\theta_{OH} = \left(\frac{K}{C_{H^+}}\right)(1 - \theta_{HC_2O_4}) \exp\left(\frac{VF}{RT}\right) \exp\left[\frac{-f(\theta)}{RT}\right] \tag{100}$$

From (98) in quasi-equilibrium $\theta_{HC_2O_4} = K'C_{ox}\theta_{OH}$, and with substitution from the previous equation and neglecting the pre-exponential coverage terms one obtains

$$f(\theta) = RT \ln K + RT \ln C_{H_2C_2O_4} - RT \ln C_{H^+} + FV \tag{101}$$

Introducing this into the rate equation and neglecting pre-exponential terms in coverage we find

$$i = kC_{H_2C_2O_4}^{\alpha} C_{H^+}^{-\alpha} \exp\left(\frac{VF}{RT}\right) \tag{102}$$

and the observed parameters are derived. Similarly, if (97) is the rate-determining step, the same argument applies and again the proper parameters are derived. Several other mechanisms have been considered (see, e.g., Ref. 226) which, however, do not give rise to the observed kinetic parameters. Johnson et al. did not favor a mechanism involving water discharge since passivation phenomena observed at about 0.9 V (N.H.E.) for the anodic oxidation of other organic compounds (presumably due to oxide formation) was not observed for oxalic acid. Other results have indicated, however, that oxalic acid oxidation does not occur on an oxide covered surface and the reaction is inhibited by the adsorption of other species such as amyl alcohol and chloride ion.[233] Thus it is not established whether the electron transfer occurs from the organic species or if water discharge is involved in the mechanism.

Thus, from a large number of mechanisms considered, Johnson et al.[226] have reduced the possibilities to the two most probable mechanisms represented by equations (91) and (92) and (91), (96), and (97). A distinction between these two possibilities cannot be made on the basis of existing experimental results.

The similarity of behavior of formic acid and oxalic acid tends to indicate a common mechanism. The proper mechanism should thus be able to account for the change in Tafel slope with temperature while the pH and concentration dependence remain unchanged.

(iii) Other Carboxylic Acids

The anodic oxidation of other carboxylic acids has either not been reported or insufficient data has been reported to enable a mechanism discussion. Acetic acid has been briefly studied and found to be unreactive at temperatures up to 100°C.[234] The electrolysis of benzoic acid has been mentioned in the literature.[235] However, no diagnostic criteria for mechanism determination have been reported. In the determination of polarization curves by a galvanostatic current scanning device for a large number of compounds, Gentile *et al.* have reported that carboxylic acids except for formic acid were not readily oxidized.[236] This study covered a number of acids including acetic, butyric, benzoic, phenyl acetic, oxalic, malonic, acrylic, maleic, glycolic, glyoxylic, and pyruvic acids. However, only the non-steady state polarization curves were obtained, and further discussion of these compounds must await experimental investigations.

3. Carboxylic Acids—Kolbe Reaction

The principal theories of the mechanism of the Kolbe electrosynthesis, as discussed earlier, have been criticized. A satisfactory summary of these has been presented by Dickinson and Wynne-Jones.[38] The main points of disagreement among these theories are, (*a*) whether the hydroxyl ion or the carboxylate ion is discharged at the anode, (*b*) the explanation of the effect of the electrode material on the reaction, and (*c*) the reason for the occurrence of the Kolbe synthesis in preference to oxygen evolution which could occur at lower anode potentials. Recent experimental investigations, particularly those of Dickinson and Wynne-Jones[38] and of Conway and Dzieciuch[37] permit a detailed analysis of the mechanism of the Kolbe reaction, in both aqueous and nonaqueous media.

(i) Overall Reactions

The dependence of Kolbe products on anode potential and electrode material has been the subject of considerable study. Only a few examples from recent studies will be discussed here since the interest is in mechanism determination and not electrosynthesis. Dickinson and Wynne-Jones[38] analyzed the composition of gases evolved from acetate solutions under various conditions and their results are presented in Table 5. They also observed that with a current density of $30 \, mA/cm^2$ in citrate and phthalate solutions, oxygen was evolved at current efficiencies of not less than 90% on the metals they examined (Pt, Ir, Pd, Au, and Ni). The results of Conway and Dzieciuch[37] with aqueous potassium formate are seen in Table 6. These results agree with those of Dickinson and Wynne-Jones, showing that the decarboxylation reaction is almost completely inhibited by the oxygen evolution process. The anodic reaction at Pd and Au in aqueous potassium trifluoro-acetate was

Table 5
The Products of Anodic Oxidation of Aqueous Acetate Solutions

Electrode material	Conc. of electrolyte, mole/liter	Current density, mA/cm²	Anode potential, V (R.H.E.)	Volume per cent of gases produced		
				CO_2	O_2	C_2H_6
Pt	1.0	1.5	2.10	19.1	73.7	0
		2.2	2.23	35.7	37.1	16.7
		14.8	2.34	55.2	2.8	34.6
		138.5	2.48	61.0	2.5	33.1
Ir	1.0	0.25	1.72	14.7	79.3	0
		0.5	2.01	31.9	42.8	17.4
		7.0	2.15	56.3	2.7	39.4
		84.2	2.30	54.6	2.2	40.6
Pt	0.1	17.0	2.16	4.6	84.3	0
		19.75	2.39	15.2	65.8	13.9
Au	1.0	30.0	1.91	5.4	90.2	0
Ni	1.0	30.0	1.77	3.2	95.1	0

found to give 87–96% oxygen, with 5–10% CO_2, and very little C_2F_6. At platinum, trifluoro-acetate gave the Kolbe coupling reaction with formation of C_2F_6 but at only moderate efficiency with about 25% oxygen evolution.[37] The coulombic efficiencies for the anodic decarboxylation of formate in anhydrous formic acid

Table 6
Coulombic Yields for CO_2 and O_2 Production in the Anodic Decarboxylation in Aqueous Potassium Formate (1 M HCOOK, H_2O, 5°C)

Electrode	Current density, A/cm²	Coulombic yields, %	
		CO_2	O_2
Pt	1×10^{-1}	93	6
	2.5×10^{-2}	72	24
	5×10^{-3}	65	33
Pd	1×10^{-1}	74	21
	2.5×10^{-2}	63	27
	5×10^{-3}	52	33
Au	2.5×10^{-2}	24	55
	3.1×10^{-3}	22	38

Table 7

Current Efficiencies for CO_2 Production in Anodic Decarboxylation in Formic Acid
(1 M HCOOK/HCOOH, 5°C)

Electrode	Current density, A/cm^2	Products	Efficiency, %
Pt	1×10^{-2}	CO_2	86
	2.5×10^{-3}	CO_2	79
	5×10^{-4}	CO_2	65
Pd	1×10^{-2}	CO_2	84
	1.2×10^{-3}	CO_2	75
	5×10^{-4}	CO_2	56
Au	1×10^{-2}	CO_2	86
	1×10^{-3}	CO_2	62
	5×10^{-4}	CO_2	55

were found to range from 55% to 86% formation of CO_2[37] (cf. Table 7). The effect of square wave frequency in the Kolbe electrosynthesis at platinum anodes has also been reported,[237] and these results are given in Table 8.

Table 8

Effect of Square Wave Frequency on Current Efficiency in the Kolbe Electrosynthesis at 25°C

	AC current density			
	0.45 A/cm^2			0.80 A/cm^2
Frequency, cps	0.5 mole/liter CH$_3$COOK 5.0 mole/liter CH$_3$COOH aqueous	0.5 mole/liter CH$_3$CH$_2$COOK 5.0 mole/liter CH$_3$CH$_2$COOH aqueous	1.0 mole/liter CH$_3$COOK in pure CH$_3$COOH	0.5 mole/liter CH$_3$COOK 7.5 mole/liter CH$_3$COOH aqueous
2	86.6*	68.3	—	—
7	85.4	48.2	92.5	—
15	85.0	40.0	92.0	88.4
25	84.3	36.3	—	—
30	81.6	32.4	92.6	88.2
69	80.7	30.8	93.3	—
80	79.2	30.2	—	—

*All values in percent.

(ii) Current–Potential Relation

Typical current–potential curves obtained with 0.2 *M* acetate solution[38] and with 1 *M* potassium trifluoroacetate[37] at platinum anodes are shown in Figures 18 and 19, respectively. The characteristic hysteresis in the *V–i* plot taken galvanostatically is shown in these figures. Current–potential behavior obtained on platinum in aqueous solution of several carboxylic acid salts is given in Figure 20, where the mean values of the Tafel constant *b* were 0.17 V for the lower section and 0.13 for the upper section.[38] The Tafel parameters obtained for the aqueous formate decarboxylation reaction are given in Table 9,[37] and for several other carboxylates in Table 10.[238] A summary of the results obtained with anhydrous formate is given in Table 11.[37] The critical current density at the inflection point below which oxygen evolution is the main anodic process (in aqueous solution) and above which the Kolbe product is formed is found to be dependent on the electrode material and on the concentration of carboxylate.[38] Davydov and D'yachko reported that with increase in the concentration of the carboxylate salt at constant anode potential, the rates of the

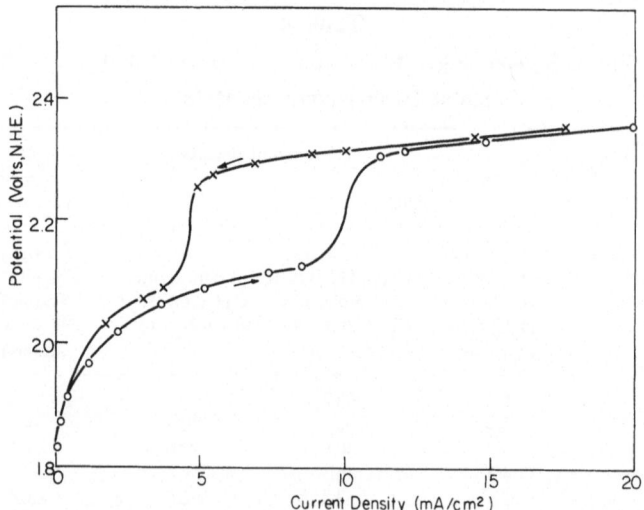

Figure 18. Hysteresis in anode potential *vs.* current density for a platinum anode in acetate solution [from Dickinson and Wynne-Jones, *Trans. Faraday Soc.* **58** (1962) 382].

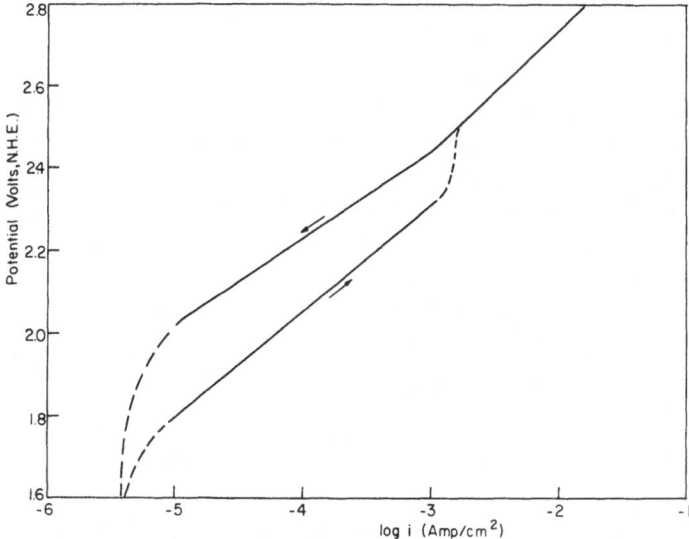

Figure 19. Current–potential curve at 5°C for the trifluoroacetate Kolbe reaction in aqueous solution at platinum [from Conway and Dzieciuch, *Can. J. Chem.* **41** (1963) 38].

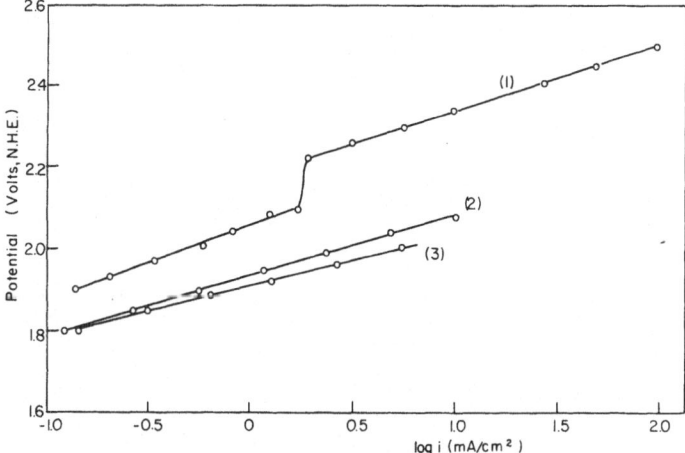

Figure 20. Anode potential *vs.* log current density on platinum for: (1) acetate, (2) citrate, (3) phthalate solution [from Dickinson and Wynne-Jones, *Trans. Faraday Soc.* **58** (1962) 382].

electrode reaction in potassium propionate and acetate solution increase along the lower section of the polarization curve and decrease along the upper section.[238] Then it appears that in the range of high anodic potentials the carboxylate may have an inhibiting effect on the rate of the anodic process.[239,240] It was further noted that excess alkali considerably increased the rate of the anodic process while excess acid had little influence.[238] However, Conway and Dzieciuch reported that addition of KOH in the aqueous formate decarboxylation reaction shifts the lower Tafel line to higher overpotentials, thus decreasing the "height" of the transition region.[37]

(iii) Other Kinetic Parameters

Kinetic parameters other than the Tafel slopes and exchange current density already given have not been determined.

Table 9
Mean Tafel Parameters for the Aqueous Formate Decarboxylation Reaction (5°C)

Electrode	Concentration	$b(\pm 10\%)$, volt	$i_0(\pm 40\%)$ A/cm^2	Current density range, A/cm^2
Pt	1 M HCOOK/H$_2$O	0.13 (lower)	1×10^{-10}	6×10^{-6}–1×10^{-2}
		0.12 (upper)	1.7×10^{-21}	2×10^{-4}–5×10^{-2}
Au	1 M HCOOK/H$_2$O	0.12 (lower)	3.9×10^{-11}	5×10^{-6}–1×10^{-3}
		0.16 (upper)	3.9×10^{-17}	1.5×10^{-4}–5×10^{-2}
Pt	1 M HCOOK + 1 M KOH/H$_2$O	0.06 (lower)	4×10^{-30}	9×10^{-6}–2×10^{-3}
		0.17 (upper)	3.2×10^{-17}	1.5×10^{-3}–1.1×10^{-1}
Pd	1 M HCOOK + 1 M KOH/H$_2$O	0.06 (lower)	4×10^{-29}	1.5×10^{-6}–5×10^{-3}
		0.12 (above hysteresis)	1.3×10^{-20}	1.5×10^{-2}–5×10^{-1}
		0.10 (in hysteresis region)	4×10^{-24}	2.5×10^{-4}–1.5×10^{-2}
Au	1 M HCOOK + 1 M KOH/H$_2$O	0.6 (lower)	8×10^{-26}	1.5×10^{-6}–2.5×10^{-3}
		0.13 (above hysteresis region)	6×10^{-26}	5×10^{-5}–5×10^{-3}
Pt	1 M HCOOK + 5 M KOH/H$_2$O	0.06 (lower)	5×10^{-31}	7×10^{-6}–3×10^{-3}
		0.16 (upper)	2×10^{-18}	5×10^{-3}–2×10^{-1}
Pd	1 M HCOOK + 5 M KOH/H$_2$O	0.06 (lower)	1×10^{-29}	2×10^{-6}–4×10^{-4}
		0.11 (upper)	2×10^{-22}	8×10^{-5}–2.5×10^{-2}
Au	1 M HCOOK + 5 M KOH/H$_2$O	0.06 (lower)	2×10^{-28}	5×10^{-6}–9×10^{-4}
		0.12 (upper)	1×10^{-19}	6×10^{-5}–5×10^{-2}

Table 10

Mean Tafel Parameters for the Aqueous Decarboxylation Reaction for Several Carboxylic Acids

Salt	Concentration, mole/liter	$b = dV/d\log i$ lower section	$b = dV/d\log i$ upper section	Transfer coefficient α lower section	Transfer coefficient α upper section
Potassium acetate	0.1	0.22	0.38	0.27	0.16
	1	0.19	0.39	0.31	0.15
	2	0.24	0.39	0.25	0.15
	3	0.24	0.39	0.25	0.15
	4	0.24	0.39	0.25	0.15
Potassium propionate	0.1	0.18	0.38	0.33	0.15
	0.5	0.22	0.44	0.27	0.13
	1.0	0.25	0.48	0.23	0.12
Potassium-n-butyrate	1.0	0.26	0.50	0.23	0.12
Potassium isobutyrate	0.1	0.24	0.56	0.25	0.11
	0.5	0.23	0.57	0.26	0.10
	1.0	0.22	0.58	0.27	0.10
Potassium n-valerate	1.0	0.28	0.66	0.21	0.09
Potassium isovalerate	0.1	0.16	0.43	0.37	0.14
	0.5	0.17	0.47	0.35	0.13
	1.0	0.18	0.51	0.33	0.12

Table 11

Mean Tafel Parameters in Anhydrous HCOOK/HCOOH, 6°C

Electrode	Conc., mole/liter	$b (\pm 10\%)$	i_0 (calc.), A/cm^2 ($\pm 40\%$)	Current density range, A/cm^2
Pt	0.1	0.14	1.2×10^{-15}	1×10^{-4}–3×10^{-2}
	1	0.14	5.2×10^{-15}	2×10^{-4}–3×10^{-2}
	5	0.12	3.2×10^{-16}	9×10^{-4}–1×10^{-1}
Pd	0.1	0.10	1.0×10^{-21}	4×10^{-5}–2×10^{-2}
		0.12	1.0×10^{-15}	1.5×10^{-5}–2.5×10^{-4}
	1	0.10	1.8×10^{-20}	5×10^{-4}–1×10^{-1}
		0.14	2.1×10^{-14}	1.5×10^{-5}–2.5×10^{-4}
	5	0.11	1×10^{-18}	5×10^{-4}–6×10^{-2}
Au	0.1	0.065	5×10^{-31}	1.8×10^{-5}–3×10^{-4}
		0.13	1.2×10^{-17}	3×10^{-4}–2.5×10^{-2}
	1	0.065	3.5×10^{-30}	2.5×10^{-5}–1×10^{-3}
		0.14	4.8×10^{-16}	1×10^{-3}–5×10^{-2}
	5	0.065	2.0×10^{-29}	5×10^{-5}–9×10^{-3}
		0.18	2.3×10^{-12}	9×10^{-3}–6×10^{-2}

(iv) *Cation Effects*

The effects observed on the Kolbe electrosynthesis when various cations are added to the electrolyte, which was attributed by Glasstone and Hickling to their action on hydrogen peroxide[241] and subsequently criticized by numerous investigators, has been recently examined in detail by Fioshin, Vasilyev, and Gaginkina.[242] The effects of 16 anions on the behavior of acetates was examined using both rotating and stationary platinum anodes. The polarization curves showed that the nature of the cation strongly affects the lower region corresponding to oxygen evolution due to discharge of water molecules, but that the effect on the upper region corresponding to acetate ion discharge was very small (cf. Figure 21). A relation was observed between the critical current density (current in the transition region) and the radius of the cation. The larger the radius of the cation, the lower was the critical current density, i.e., the lower the potential for the inhibition of oxygen

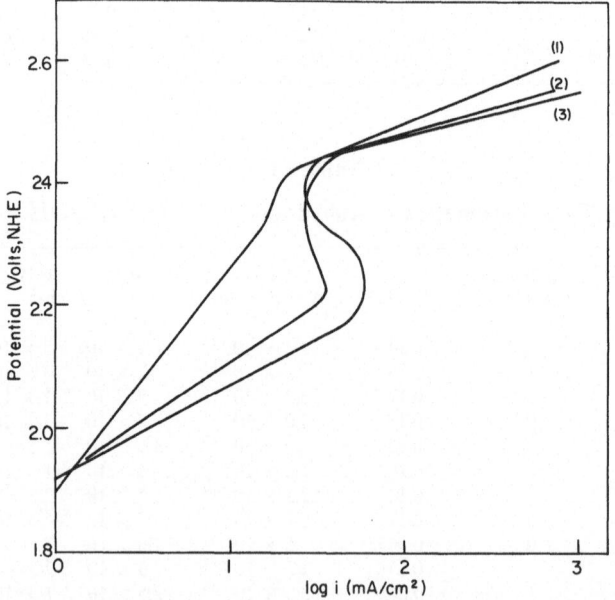

Figure 21. Polarization curves for 0.1 N acetates of the alkali metals: (1) CH$_3$COOK, (2) CH$_3$COONa, (3) CH$_3$COOLi [from Fioshin, Vasilyev, and Gaginkina, *Dokl. Akad. Nauk. SSSR* **135** (1960) 909].

evolution and the discharge of acetate ions. The nature of the cation effects was also dependent on the concentration as expressed by $I_{critical} = k/C^n$. The results of maximum yield of the Kolbe product are compared with the radius of cation for several cations in Table 12. Attainment of the maximum yield corresponds to the largest drop in the critical current density, i.e., to the strongest suppression of the oxygen evolution reaction.

The effect of the nature of the cation on the anodic reaction was assumed to be connected with an adsorption of the cation in the double layer. The critical current density was attributed to inhibition of oxygen evolution. It was observed that the critical current was lowered only when the radius of the added cation was larger than that of the cation of the acetate constituting the base solution. The introduction of a cation with a larger radius and, hence, with a larger adsorbability on a given electrode, leads to its preferential adsorption on the electrode. Dickinson and Wynne-Jones also support the view that added ions reduce the activity of acetate ions in the double layer.

(v) Coverage

The extent of film formation in the formate reaction was examined using cathodic galvanostatic discharge.[37] Arrests in the

Table 12
Dependence of the Yield of CO_2 and C_2H_6 upon the Nature of the Cation

Cation	I_{max} at $C = 0.1 N$ A	n $(i_{cr} = kc^{-n})$	Current yield of C_2H_6, %	Radius of the cation, Å
Li^+	46.8	0.63	64.0	0.60
Na^+	38.9	0.63	66.2	0.93
K^+	22.4	0.49	68.2	1.33
NH_4^+	17.35	0.42	68.7	1.43
Rb^+	15.85	0.380	66.4	1.48
Cs^+	13.5	0.280	—	1.99
$(CH_3)_4N^+$	11.2	0.205	76.2	—
Mg^{2+}	61.5	0.54	—	0.65
Ca^{2+}	51.1	0.56	61.1	0.99
Sr^{2+}	38.9	0.515	61.7	1.13
Ba^{2+}	30.9	0.46	68.3	1.35

galvanostatic transients up to 2,000 $\mu C/cm^2$, indicated films up to 20 monolayers per apparent cm^2. For the trifluoroacetate reaction in anhydrous trifluoroacetic acid, a build up of an adsorbed layer with time was observed; however, this did not exceed the limit for one monolayer on Pt. Based on these results and examination of adsorbed intermediates using potential decay measurements, the significance of the transition region was discussed by Conway and Dzieciuch. They suggest that the transition from lower current density behavior to that at high current density, associated with a sharp change of potential, must be ascribed to filling up of the surface by the adsorbed species. The transition behavior, which shows the characteristics of passivation phenomena, are thus ascribed to formation of a monolayer of adsorbed reaction intermediates.

(vi) Mechanistic Conclusions

The formation of a film at a limiting current density in aqueous solutions, is generally regarded as a necessary occurrence before the Kolbe reaction proceeds at appreciable current efficiencies. In nonaqueous solutions film formation still occurs, but it does not seem to be a prerequisite, since some Kolbe products were found below the critical current density.

The discussion below must be separated for nonaqueous and aqueous conditions since it is not expected that similar mechanisms would be obtained in both systems.

In aqueous solutions there will exist a competition for discharge between the carboxylate ions and water (or hydroxyl ion). At current densities below the critical value, where oxygen is the principal anodic product, the discharge of water (or OH^-) will predominate; however, a steady-state concentration of carboxylate radical will be present on the electrode surface. This coverage by carboxylate radicals will depend on the current density, and as it increases the discharge of water will become increasingly difficult, [38,243-245] and discharge onto an occupied site is considered very difficult. Eventually a current density is obtained such that the electrode is completely covered by carboxylate radicals and further discharge must occur on already occupied sites. Thus the critical current density is the current at which removal of adsorbed carboxylate radicals by reaction with hydroxyl radical or water molecules becomes less than the rate of discharge. The increase in potential is then associated with the difficulty of discharging a carboxylate ion onto an occupied site. This argument implies[38] that whereas an adsorbed carboxylate radical is a fairly stable species, the presence of two acetate radicals on a single site causes instability, and degradation occurs, giving CO_2 and the Kolbe product. The difference between values of the critical current density obtained on different electrodes has been attributed to a dependence of the relative rates of the discharge of the carboxylate ion and water (or OH^-) or of the rates of the removal reactions.[38] For example, it can be supposed that the rate of discharge of water is more rapid on Au than on Pt, and, consequently, on Au the carboxylate radicals never reach a sufficient coverage for the Kolbe reaction to occur. It has also been suggested that for "chemical reaction," the oxide films are more stable and thicker at Au and Pd than at Pt and cause inhibition of carboxylate ion discharge.[37]

The hysteresis effect in the anode potential vs. current density curves is attributed to the fact that water (or OH^-) does not easily discharge on occupied sites. With the experimental evidence discussed, Conway and Dzieciuch have concluded for the aqueous formate reaction that the lower slope region corresponds to the rate-determining step

$$HCOO \cdot \rightarrow H \cdot + CO_2 \qquad (87)$$

associated with the unimolecular decomposition slope of $2.3\, RT/F$. The upper region is then associated with the discharge of formate radicals at high coverage

$$HCOO \cdot \rightarrow CO_2 + H^+ + e \qquad (66)$$

or discharge of further formate ions onto the layers or film of adsorbed formate. Tafel slopes of greater than $2RT/F$ are explained in terms of heavier layer films[37,238] which have been discussed for processes on anodic oxide films.[246–249] Under such conditions, the slope is proportional to the thickness of the film and the transfer coefficient for a single-electron charge-transfer step becomes $\alpha = \delta/t$ where δ is the half-jump distance or barrier half-width, and t is the film thickness.[37]

From studies in nonaqueous formic acid solutions, Conway

and Dzieciuch have proposed the following scheme for formate decarboxylation:

$$HCOO^- \rightarrow HCOO^-_{ad} + e \qquad (103)$$

$$HCOO^-_{ad} \rightarrow CO_2 + H_{ad} \qquad (104)$$

$$H_{ad} \rightarrow H^+ + e \qquad (62)$$

or

$$HCOO^- \rightarrow HCOO^-_{ad} + e \qquad (103)$$

$$HCOO^-_{ad} \rightarrow CO_2 + H^+ + e \qquad (66)$$

For Pt or Pd, the results are consistent with step (104) rate-determining under Temkin conditions, or with step (66) rate-determining at full coverage. Further distinction between these schemes was not possible.

4. Aldehydes and Ketones

No mechanism studies are available for the anodic oxidation of ketones and very few for aldehydes, the latter being carried out primarily in connection with the oxidation of alcohols and carboxylic acids. The anodic oxidation of formaldehyde and acetaldehyde will be only briefly discussed.

(i) Formaldehyde

The anodic oxidation of formaldehyde in acid media yields CO_2 at essentially 100% current efficiency.[77,207] From controlled potential oxidation experiments, it was reported that the amount of formaldehyde which is oxidized is independent of the anode potential between 0.4 and 1.0 V (N.H.E.).[206] In basic solutions the main reaction product is formate which is only slowly oxidized in alkaline media.[207,250]

Studies of the oxidation of formaldehyde have been carried out primarily with nonstationary techniques, i.e., cyclic voltametry and chronopotentiometry. The polarograms obtained with formaldehyde show evidence of two oxidation peaks,[99,206,207] which are shifted depending on the pH of the electrolyte. The peak currents which are not diffusion controlled, were found to be larger for formaldehyde than for methanol or formic acid.[207] Plots of $\log i$ vs. V from the cyclic voltametric data for the oxidation of 0.078–0.13 M formaldehyde in 0.1 M NaOH at platinized nickel anodes gave Tafel slopes on the order of $2RT/F$.[207] This Tafel slope ($2RT/F$) was also reported from cyclic voltametric measurements with a rotating platinum electrode.[180] With 0.05 M formaldehyde in 1 M H$_2$SO$_4$, Tafel slopes of approximately $3RT/2F$ were reported,[207] and in 5 M H$_2$SO$_4$, $RT/2F$.[207] The reaction order with respect to formaldehyde was determined as approximately 0.5 in one study[180] and approximately 1 in another investigation.[206] The reaction is reported to be first order with respect to the electrode surface area[206] and of fractional order (approximately 0.5) with respect to the pH.

With the apparent disagreement of experimental results and the lack of any steady-state measurements, it is difficult to suggest a proper mechanism although several have been proposed. Bagotzky and Vasilyev[180] proposed dissociation of the adsorbed organic molecule as the rate-determining step, i.e.,

$$CH_2O \rightleftharpoons CH_2O_{ad} \tag{105}$$

$$CH_2O \xrightarrow{r.d.s} CHO_{ad} + H_{ad} \tag{106}$$

followed by oxidation of the adsorbed CHO· radical to CO$_2$, probably through a formic acid radical intermediate HCOO.

This type of mechanism is subject to the criticisms already given under the discussion of formic acid (cf. p. 110). Liang and Franklin[206] also suggested that the slow step is a process involving adsorption of the formaldehyde, but they do not give any argument to support this mechanism. Buck and Griffith[207] have proposed the following scheme for reaction of the hydrated formaldehyde (methylene glycol) in alkaline solution, with the last step rate-determining:

$$CH_2(OH)_2 + OH^- \rightleftharpoons CH_2(OH)O^- + H_2O \tag{107}$$

$$CH_2(OH)O^- \rightarrow CH(OH)O_{ad}^- + H_{ad} \tag{108}$$

$$H_{ad} + OH^- \rightarrow H_2O + e \tag{109}$$

$$CH(OH)O_{ad}^- + OH^- \xrightarrow{r.d.s} HCOO^- + H_2O + e \tag{110}$$

Under Langmuir conditions, a rate equation of the following form is derived:

$$i = k' C_{CH_2(OH)_2}(1 - \theta)C^2_{OH^-} \exp\left[(1 + \alpha)\frac{VF}{RT}\right] \quad (111)$$

where k' includes equilibrium constants of the previous reactions assumed in equilibrium, and θ is the total electrode coverage. A Tafel slope of $2RT/3F$ is thus predicted for this mechanism, and a reaction first order in the reactant methylene glycol, first order in the free electrode surface, and second order in the OH^- concentration. These parameters do not agree with the available experimental data. It is concluded that since the proposed mechanisms do not agree with present experimental evidence, further experimental study must be carried out before an acceptable mechanism can be proposed.

(ii) *Acetaldehyde*

The principal studies of acetaldehyde oxidation are those of Bogdanovskii and Shlygin,[224,225,251] who have proposed an electron-radical mechanism which is of the form:

$$CH_3CHO \rightarrow [CH_3CHO]^+ + e \quad (112)$$

$$[CH_3CHO]^+ + H_2O \rightarrow CH_3CO_{ad} + H_3O^+ \quad (113)$$

$$2H_2O \rightarrow H_3O^+ + OH_{ad} + e \quad (114)$$

$$\overset{\displaystyle O}{\overset{\displaystyle \|}{CH_3C_{ad}}} + OH_{ad} \rightarrow \overset{\displaystyle O}{\overset{\displaystyle \|}{CH_3COH}} \quad (115)$$

with acetic acid, which is only slowly oxidized, being the primary product. The slow step is taken to be formation of the hydroxyl radical, and the reaction is terminated when the hydroxyl radical reacts to form an oxide (or adsorbed oxygen) which covers the electrode surface. The experimental data for the establishment of the mechanism is not available.

5. Alcohols

Although the electrolysis of alcohols was reported in 1881,[252] their electrochemical behavior was not seriously studied until the recent interest in the development of low temperature alcohol fuel cells. Many of the investigators who have studied formic acid have also included alcohol in their investigations and some similarity in behavior has been indicated.

(i) Methanol

Müller and his co-workers made the first mechanistic investigation of methanol, determining current–potential curves on several noble metals.[26] In 1954, Pavela[77] studied methanol oxidation using galvanostatic steady-state methods. Recent investigations have been carried out almost exclusively with non-steady state methods.[102,164,180,206,207,212,215–217,253–255] One study has recently appeared on the oxidation of methanol at Pt anodes, apparently using steady-state galvanostatic methods.[256]

(a) Overall reaction—The anodic oxidation of methanol at platinum electrodes has been found to proceed through the intermediate formation of formaldehyde and formic acid with the final product depending on pH of the solution. In acid solution nearly complete oxidation to CO_2, according to the reaction

$$CH_3OH + H_2O \rightarrow CO_2 + 6H^+ + 6e \qquad (116)$$

has been established with traces of HCOOH and HCHO being detected.[26,52,53,77,207,212,216] In alkaline solution the principal product is formate, with lesser amounts of formaldehyde and some carbonate being formed. The overall scheme at platinum anodes may be written

$$CH_3OH \xrightarrow{-2e} HCHO \xrightarrow{-2e} HCOOH \xrightarrow{-2e} CO_2 \qquad (117)$$

with formic acid being oxidized in acid solutions, but with formate ion being unreactive in alkaline solution.

In alkaline solution at Pd anodes, the overall reaction is given as

$$CH_3OH + 8OH^- \rightarrow CO_3^{--} + 6H_2O + 6e \qquad (118)$$

with a current efficiency for carbonate production greater than 80%.[256]

(b) Open-circuit potentials—Experimental observations indicate the same type of open circuit potential behavior for methanol that was observed for formic acid.[212] (cf. p. 103). The delay time observed in the potential decay curve was of the same order of magnitude. However, a similar relation between τ and the methanol concentration was not observed. The decaying potentials pass through a minimum and reach steady values in the region of 0.2–0.3 V (N.H.E.) in 1 N H_2SO_4, and -0.54 V (N.H.E.) in 1 N NaOH. Pavela reported a decrease in open-circuit potential with increasing temperature and a concentration dependence $dE_R/d \log C_{CH_3OH}$ = -40 mV, in alkaline solution. The pH dependence of the rest potential is $dE_R/d_{pH} \approx -RT/F$. These results indicate that the open-circuit potential for methanol is established by a mechanism similar to those discussed for formic acid.

(c) *Current–potential relations*—Current–potential curves have been determined by the galvanostatic steady-state method on platinized Pt anodes at 25°C.[77] In 1 N NaOH, a Tafel region was observed in the potential range from -0.35 to $+0.15$ V (N.H.E.), with a slope of 0.12 V ($2RT/F$). A limiting current was obtained above 0.15 V. In 1 N H_2SO_4, a Tafel region from 0.5 to 0.7 V (N.H.E.) was reported with a slope of 0.08 V ($\sim RT/F$). Above 0.7 V, a region of potential oscillation was observed, and above this a limiting current. Pavela stated that the Tafel lines determined were not well reproducible. A Tafel slope of $2RT/F$ has been reported for studies of methanol oxidation at Pd anodes.[256]

The Tafel regions have also been observed in cyclic voltametric measurements. Buck and Griffith found that the ascending part of the current peak obtained in the anodic sweep obeyed the Tafel equation with a slope of 0.21–0.28 V in the pH range 10–14.[207] Bagotzky reported a Tafel slope of $2RT/F$ over the whole pH range for both the ascending and descending portion of all three current peaks obtained in the anodic sweep.[180]

(d) *Other kinetic parameters*—The pH dependence and concentration dependence in the anodic oxidation of methanol have only been determined from cyclic voltametric measurements. The parameters which have been determined are summarized in Table 13.

(e) *Mechanistic considerations*—Several types of mechanisms have

Table 13
Kinetic Parameters for Methanol Oxidation

T	b	C_{CH_3OH}, mole/liter	pH	i_0, A/cm^2	$\left(\dfrac{d\log i}{d\,\mathrm{pH}}\right)_v$	$\left(\dfrac{d\log i}{d\log C_{CH_3OH}}\right)_v$	Ref.
15°	$2RT/F$	0.5	14	10^{-7}	—	0	77
23°	~ 0.25 V	0.1	10–14	5×10^{-8}	~ 0.5	~ 1	207
25°	$\sim RT/F$	—	0.3	—	—	0	77
25°	$2RT/F$	10	0.4	10^{-8}	~ 0.5	~ 0.5	180
30°	$\sim RT/F$	1	~ 0	10^{-11}	—	—	254
—	—	—	—	—	—	1	206
25°	$2RT/F$	0.2–5	~ 14	$\sim 10^{-5}$	~ 0.67	~ 0.33	256

been suggested for the anodic oxidation of methanol. Bagotzky observed similar kinetic parameters for the anodic oxidation of several organic species and thus proposed the dissociative mechanism as a general scheme for organic oxidation[180] (cf. p. 107). The remarks concerning this mechanism which were made in the discussion for formic acid apply equally here and thus will not be further discussed. A second mechanism was proposed by Pavela[77] and later supported in a similar form by Buck and Griffith.[207] The mechanisms, proposed for alkaline solution, may be written

$$CH_3OH \rightleftharpoons CH_3OH_{ad} \tag{119}$$

$$OH^- \rightleftharpoons OH_{ad}^- \tag{120}$$

$$CH_3OH_{ad} \rightarrow CH_3O_{ad} + H^+ + e \tag{121}$$

$$CH_3O_{ad} \rightarrow CH_2O_{ad} + H^+ + e \tag{122}$$

$$CH_2O_{ad} + OH_{ad}^- \rightarrow CH_2(OH)O_{ad}^- \tag{123}$$

$$CH_2(OH)O_{ad}^- \rightarrow CH_2(OH)O_{ad} + e \tag{124}$$

$$CH_2(OH)O_{ad} \rightarrow HCOOH + H^+ + e \tag{125}$$

$$OH^- + H^+ \rightleftharpoons H_2O \tag{126}$$

Pavela did not indicate a rate-determining step. However, he thought that any of the steps involving electron transfer could be rate limiting. From his experimental observation of Langmuir-type adsorption and a Tafel slope of $2RT/F$, the first electron transfer (121) is indicated as the rate-determining step. This would lead to a first-order dependence on methanol concentration which has been reported in some investigations; however, a pH dependence is not explained. The large region of potential independence of coverage suggests that Temkin conditions need not be considered.

None of the mechanisms suggested for the anodic oxidation of methanol can explain all of the observed kinetic parameters and a definite mechanism cannot at present be established.

(ii) Ethanol

(a) *Overall reaction*—The overall reaction for the anodic oxidation of ethanol has been established by product distribution studies employing a gas chromatograph for the product analysis.[257] At 0.75 V (N.H.E.) in 1 N H_2SO_4 solution, the primary product was acetaldehyde with only small amounts of acetic acid. Acetic acid and acetaldehyde accounted for 100% of the products based on the quantity of current used in the experiment. The overall anodic process can thus be written as

$$CH_3CH_2OH \rightarrow CH_3CHO + 2H^+ + 2e \qquad (127)$$

and

$$CH_3CH_2OH + H_2O \rightarrow CH_3COOH + 4H^+ + 4e \qquad (128)$$

with (127) being the predominant reaction. It has also been reported that the acetic acid can be obtained at a current efficiency of 90% under special conditions.[258]

(b) Open-circuit potential—Frumkin and Podlovchenko have reported a study of the nature of potentials of the platinum electrode arising in ethanol solution.[213,214] They found that the steady-state open-circuit potential [$+0.16$ V (N.H.E.)] did not depend on the starting potential as long as this exceeded a limiting value of 0.13 V. From high potentials (e.g., 1 V) the open-circuit potential rapidly decreased to the steady value (within ~ 2 min). The rest potential was found to depend on pH in the normal way, i.e., $(dE_R/d \text{ pH}) = -RT/F$. The concentration dependence of the rest potential was not indicated.

The mechanism proposed for establishment of the open-circuit potential has been discussed in the section on formic acid (cf. p. 103), and will not be repeated here.

(c) Current–potential relations—Tafel lines have been reported for ethanol oxidation in $1 N$ H_2SO_4 and $1 N$ K_2SO_4 on platinized Pt in the potential region 0.4–0.55 V (N.H.E.), with slopes of $2RT/F$.[213] Current–potential curves determined by a "quasi-steady-state" method corresponding to an average potential sweep rate of 0.1 V/min and also by cyclic voltametry with a sweep rate of 0.025 V/min exhibited linear Tafel behavior in the potential range 0.55–0.75 V (N.H.E.), with slopes of 0.11 V.[257]

(d) Other kinetic parameters—Extrapolation of linear Tafel lines to the reversible potential calculated from thermodynamic data for reaction (127) ($E^0 \approx 0.17$ V) gives exchange currents of about 5×10^{-7} A/cm^2. The pH dependence as indicated by a shift of

Tafel line with pH is $(d \log i/d \text{ pH})_V \approx 0.5$. The concentration dependence obtained by the change of height of the current peaks with concentration in cyclic voltametric measurements is given as $(d \log i/d \log C_{CH_3CH_2OH})_V \approx 0.5$.[180]

(e) *Mechanistic conclusion*—Frumkin and Podlovchenko have suggested the dissociative mechanism which has been previously discussed. Rightmire et al.[257] proposed a fast chemisorption step involving charge transfer followed by a rate-limiting electrochemical step:

$$CH_3CH_2OH \rightleftharpoons CH_3\dot{C}HOH_{ad} + H^+ + e \qquad (129)$$

$$CH_3\dot{C}HOH_{ad} \xrightarrow{\text{r.d.s.}} CH_3CHO + H^+ + e \qquad (130)$$

For Langmuir conditions the rate equation can be written as

$$i = k\theta_E \exp\left(\frac{\beta V F}{RT}\right) \qquad (131)$$

where θ_E is the partial surface coverage by adsorbed ethanol. From equilibrium in step (129) one obtains

$$\theta_E = Kc_E c_{H^+}^{-1}(1 - \theta_E) \exp\left(\frac{VF}{RT}\right) \qquad (132)$$

Combining equations (131) and (132) one has

$$i = kc_E c_{H^+}^{-1} \exp\left[(1 + \beta)\frac{VF}{RT}\right] \qquad (133)$$

None of the parameters derived from this expression agree with the experimental observations.

Under Temkin conditions the rate equation may be written as

$$i = k\theta_E \exp\left(\frac{\beta F V}{RT}\right) \exp\left[\frac{\alpha f(\theta)}{RT}\right] \qquad (134)$$

For equilibrium in (129), and assuming that $\theta_E \sim \theta_T$ and the pre-exponential terms in coverage are negligible (this is reasonable for $0.2 < \theta < 0.8$), one finds that

$$f(\theta) = \ln K + \ln C_E - \ln C_{H^+} + \frac{FV}{RT} \qquad (135)$$

and on substitution, again neglecting the pre-exponential coverage term and assuming $\alpha \doteq \beta \doteq 0.5$, one obtains

$$i = K' C_E^\alpha C_{H^+}^\alpha \exp\left(\frac{VF}{RT}\right) \tag{136}$$

The concentration and pH dependence are now as observed; however, the proper Tafel slope is not derived. It may be concluded that a mechanism which agrees with the experimental observation has not as yet been established.

(iii) Other Alcohols

Some investigations of other alcohols have been carried out, e.g., the anodic oxidation of n-propanol[259] and of phenol;[260] however, the results are not sufficient to permit a mechanistic discussion.

6. Hydrocarbons

The study of electrochemical oxidation of hydrocarbons, except for a few compounds like benzene, was begun only very recently (primarily beginning about 1960) and as yet little has been published in this area. Several studies have been made of the performance of various hydrocarbons in fuel cells with determination of current efficiency and product analysis (cf. introduction); however, kinetic parameters were not determined. The few systems which have been studied from a proper mechanistic point of view will be discussed here.

(i) Alkynes—Acetylene

The mechanism of the electro-oxidation of acetylene has been studied on platinum electrodes by Johnson, Wroblowa, and Bockris and this discussion will essentially follow their paper.[193]

(a) *Overall reaction*—The overall anodic reaction has been established throughout the pH range to be

$$C_2H_2 + 4H_2O \rightarrow 2CO_2 + 10H^+ + 10e \tag{137}$$

with virtually 100% conversion to CO_2 in the Tafel region of potentials.

(b) *Open-circuit potentials*—The reversible potential, calculated from thermodynamic data for the above reaction is -0.05 V at $80°C$. The observed open-circuit potential for acetylene is 0.26 V in $1 N$ H_2SO_4 and -0.58 V in $1 N$ NaOH, being attained immediately upon introduction of the acetylene and remaining constant for periods of several days. The rest potential was found to be linearly dependent on pH with dE_R/d pH $\approx -RT/F$, and independent of the partial pressure of acetylene. It was established independently of the temperature and purification of the system and stirring had no effect. Interpretations of the open-circuit potential have been discussed and various suggestions have been made including various types of mixed potentials and the reversible equilibrium potential for the water discharge reaction. No definite mechanism could be established.

(c) *Current–potential relations*—The experimental measurements were made using steady-state potentiostatic and galvanostatic methods. A linear Tafel region was observed in the potential region 0.6–0.9 V (N.H.E.) ($1 N$ H_2SO_4) with a slope of RT/F. Above 0.9 V the slope rapidly increased until passivation occurred (~ 1.0 V) with the current decreasing to negligible values. Extrapolation of the Tafel line to the calculated reversible potential gave an exchange current density of about 5×10^{-18} A/cm^2 (geom).

(d) *Activation energy*—Calculation of apparent energies of activation from the variation of current with temperature at constant potential gave 21.5 ± 1 kcal/mole in $1 N$ H_2SO_4 and 26 ± 1 kcal/mole in $1 N$ NaOH. No significance was attached to the difference in acidic and basic solutions.

(e) *Other kinetic parameters*—The Tafel lines determined with solutions of different pH were shifted by about 50 mV per pH unit, i.e., $(dV/d$ pH$)_i \sim -50$ mV, with $(d \log i/d$ pH$)_V = 0.8$. The steady-state current was found to decrease with increasing partial pressure of acetylene, $(d \log i/d P_{C_2H_2})_V < 0$.

(f) *Mechanistic conclusions*—The observed pressure dependence indicates that the rate-determining step in the reaction involves a species other than acetylene. Under Langmuir conditions, the Tafel slope of RT/F indicates a chemical reaction following the first charge transfer step for low coverage, or following any charge transfer step at full coverage. For full coverage no pressure dependence would be expected. Thus, the following mechanism has been suggested:

$$C_2H_2 \rightleftharpoons C_2H_{2\,ad} \tag{138}$$

$$H_2O \rightleftharpoons \cdot OH_{ad} + H^+ + e \tag{78}$$

$$C_2H_{2\,ad} + \cdot OH_{ad} \xrightarrow{\text{r.d.s.}} \text{intermediate} \tag{139}$$

$$\text{intermediate} \overset{H_2O}{\rightleftharpoons} 2CO_2 + 9H^+ + 9e \tag{140}$$

The discharge of water is proposed over the whole pH range since the experimental results indicate no change of parameters from acid to alkaline media and OH^- species would not be present in acidic solutions in sufficient quantity to support the observed currents. For Langmuir conditions, the rate of reaction for the proposed mechanism may be written as

$$i = K\theta_{C_2H_2}\theta_{OH} \tag{141}$$

and from equilibrium in the previous steps,

$$\theta_{OH} = K(1 - \theta_T)a_{H^+}^{-1}\exp\left(\frac{VF}{RT}\right) \tag{142}$$

which gives on substitution:

$$i = K'\theta_{C_2H_2}(1 - \theta_T)a_{H^+}^{-1}\exp\left(\frac{VF}{RT}\right) \tag{143}$$

The coverage by OH radicals will be low in the Tafel region, and since the steps after the rate-determining step are not in equilibrium, as a result of removal of the product CO_2, the total coverage may be approximated $\theta_T \approx \theta_{C_2H_2}$. Thus, if $\theta_{C_2H_2} > 0.5$, a negative pressure dependence is predicted. The other derived parameters compare favorably with experimental results with $(d\log i/d\,pH)_V = 1$ and $(dV/d\,pH)_i = -70\,mV$.

(ii) Alkenes

Several studies have been conducted with olefinic hydro carbons in fuel cells. The performance of ethylene and other hydrocarbons has been determined in terms of the current–potential curves obtained under load for the hydrocarbon/oxygen fuel cells. Product studies were also carried out in these systems; however, no discussion of mechanisms was attempted (see, e.g., Ref. 261). The anodic oxidation of ethylene has been investigated by several authors,[32,64,69,217] and other olefins have been examined.[33,68,196]

(a) Overall reaction—The overall reaction for the anodic oxidation of ethylene on Pt electrodes in alkaline solution has been established, by determination of the amount of ethylene consumed in the reaction, to be the following:[32,69]

$$C_2H_4 + 16OH^- \rightarrow 2CO_3^= \rightarrow 10H_2O + 12e \qquad (144)$$

Studies in the acidic media also indicate complete conversion to CO_2 according to the equation[32]

$$C_2H_4 + 4H_2O \rightarrow 2CO_2 + 12H^+ + 12e \qquad (145)$$

Earlier studies had indicated aldehyde and ketone formation in the oxidation of ethylene;[57] however, later work using mass spectrographic analysis and gas chromatography reported no products other than CO_2 and water.[53,58] The coulombic efficiency determinations for the anodic oxidation of other unsaturated hydrocarbons in acid solution indicate that CO_2 is the main reaction product[23] (cf. Table 17). For lower olefins, i.e., C_2H_4 and C_3H_6, oxidation is complete. For larger compounds, other products are formed; however, the primary product is CO_2. A general equation for hydrocarbon electro-oxidation can thus be written as

$$C_xH_y + 2xH_2O \rightarrow xCO_2 + (4x + y)H^+ + (4x + y)e \qquad (146)$$

The charge passed through the electrolytic cell per mole of CO_2 produced during complete oxidation in acidic medium is then $(4x + y)/x$ F. This value will be 6 for olefinic and a fraction approaching 6 for other hydrocarbons, i.e., greater than 6 for saturated compounds and less than 6 for compounds with greater unsaturation than olefins. The reversible potentials are calculated assuming complete oxidation.

(b) Open-circuit potential—The open-circuit potentials for ethylene in 1 N NaOH were reported as -0.56 V (N.H.E.), being achieved rapidly (<1 min) upon introduction of the gas to the cell.[32,69] A minimum in the open-circuit potential, as reported for other organic compounds, i.e., HCOOH, was not observed with unsaturated hydrocarbons[32,33] on platinized Pt electrodes, i.e., the open-circuit potential decreased to a steady value which was not followed by a subsequent increase in potential. A minimum has been observed, however, by some investigators, suggesting that it may depend on the pretreatment of the electrode, e.g., it may involve absorbed hydrogen.

The open-circuit potentials were found to decrease by approximately RT/F per unit of increasing pH, but to be essentially independent of temperature (within ± 5 mV) over the range 25°–80°C and of hydrocarbon partial pressure (within ± 5 mV) over the range 1 to 10^{-4} atm. Values of the open-circuit potential obtained with platinized platinum in 1 N H_2SO_4 are given in Table 14.

(c) *Current–potential behavior*—Current–potential curves have been reported for unsaturated hydrocarbons determined by the methods of cyclic voltametry,[64,68] voltage step,[217] and potential sweep voltametry,[217,262] and by steady-state galvanostatic and potentiostatic methods.[32,33,69] For ethylene oxidation at platinum electrodes in 1 N H_2SO_4, two current peaks, one at about 0.84 V and the other at about 0.99 V, were obtained during an anodic potential sweep.[64] Tafel slopes determined by non-steady state methods for ethylene oxidation at 80°C in 1 M H_2SO_4 were in the

Table 14

Open Circuit Potentials in the Presence of Unsaturated Hydrocarbons

Species dissolved in 1 N H_2SO_4	Open-circuit potential, V (N.H.E. at 80°C)
N_2	0.23
C_2H_4	0.25
C_3H_4	0.29
C_3H_6	0.26
$C_4H_8 - 1$	0.25
$C_4H_8 - 2$	0.24
C_4H_6	0.25
C_6H_8	0.15
C_6H_6	0.37

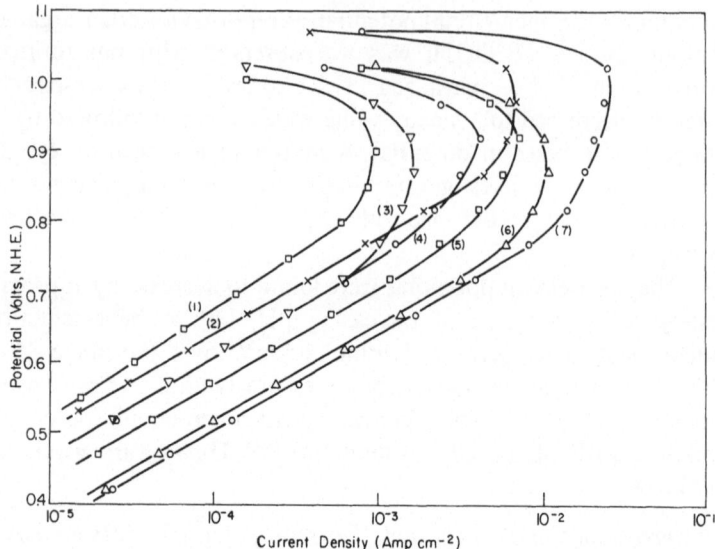

Figure 22. Current–potential relations for the anodic oxidation of unsaturated hydrocarbons at platinum, $1 N$ H_2SO_4, 80°C: (1) benzene, (2) butadiene, (3) 2-butene, (4) 1-butene, (5) propylene, (6) ethylene, (7) allene [from Bockris, Wroblowa, Gileadi, and Piersma, *Trans. Faraday Soc.* **61** (1965) 2531].

range 0.22 to 0.29 V. These results are not relevant for the discussion of the mechanism under steady-state conditions. Figure 22 shows a comparison of the current–potential curves determined potentiostatically at platinized Pt anodes in $1 N$ H_2SO_4 for several hydrocarbons. The characteristic regions have been previously discussed by Wroblowa, Piersma, and Bockris.[32,33,104] In general, a linear Tafel region with a slope of $2RT/F$ over 3 decades of current is obtained. The apparent passivation above about 0.9 V (N.H.E.) has been attributed to adsorbed oxygen or surface oxide formation. There are indications that at higher temperatures in phosphoric acid, the Tafel slope approaches RT/F.[262]

(*d*) *Exchange current densities*—The exchange current densities were estimated by extrapolation of the linear Tafel regions of the current–potential relations to the calculated reversible potentials. The values of i_0 for hydrocarbon oxidation in $1 N$ H_2SO_4 at 80°C are of the order of 10^{-8} A/cm² (geometric area) for platinized platinum anodes and 10^{-10} A/cm² for bright platinum electrodes.

Table 15

Reaction Rates and Reversible Potentials for Unsaturated Hydrocarbons

Hydrocarbon	Reversible potentials, V (R.H.E.)	i_0 (A/cm^2)	$i_{p.z.c.}$ (A/cm^2)
C_2H_4	0.06	5.0×10^{-8}	5.2×10^{-5}
C_3H_4	−0.01	2.1×10^{-8}	6.4×10^{-5}
C_3H_6	0.07	2.0×10^{-8}	2.0×10^{-5}
C_4H_6	0.04	0.5×10^{-8}	0.8×10^{-5}
$C_4H_8 - 1$	0.07	1.2×10^{-8}	1.4×10^{-5}
$C_4H_8 - 2$	0.07	1.2×10^{-8}	1.4×10^{-5}
C_6H_6	0.07	0.2×10^{-8}	0.6×10^{-5}

The i_0 values for the hydrocarbons examined are given in Table 15, and are seen to vary from ethylene to benzene within about an order of magnitude. For the comparison of relative rates of oxidation, it is perhaps more relevant to compare the current densities at a fixed potential, in particular, the potential of zero charge. The current densities obtained by extrapolation of the Tafel regions to 0.5 V, the approximate p.z.c. of Pt,[172] are also given in Table 15.

(e) *Activation energies*—The apparent energies of activation for the anodic oxidation reactions were calculated from the slope of the log i vs. $1/T$ plots taken at constant overpotential, and are given in Table 16.

Table 16

Apparent Energies of Activation for the Electrochemical Oxidation of Unsaturated Hydrocarbons

Hydrocarbon	$\Delta H_\eta^{\ddagger}$ kcal/mole average	$\eta(V)$ average	$\Delta H_{\eta=0}^{\ddagger}$ kcal/mole theoretical
C_2H_4	21.0	0.46	26.3
C_3H_4	19.7	0.62	26.8
C_3H_6	23.7	0.55	30.0
C_4H_6	20.1	0.63	27.3
$C_4H_8 - 1$	22.3	0.55	28.6
$C_4H_8 - 2$	23.0	0.55	29.3
C_6H_6	22.9	0.70	30.9

Extrapolation to $\eta = 0$ could not be made experimentally because of the narrow range of potential over which measurements could be taken, particularly at lower temperatures. A theoretical extrapolation can be made if one considers a rate equation of the form

$$i = k \exp\left(\frac{-\Delta H_{\eta=0}^{\ddagger}}{RT}\right) \exp\left(\frac{\alpha \eta F}{RT}\right) \tag{147}$$

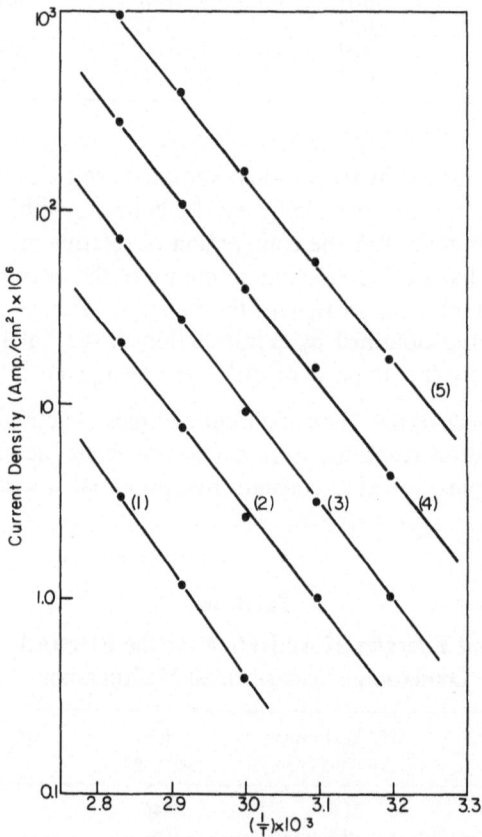

Figure 23. Temperature dependence in anodic oxidation of propylene at platinum in $1\,N$ H_2SO_4: (1) 0.42 V, (2) 0.52 V, (3) 0.62 V, (4) 0.72 V, (5) 0.82 V [from Piersma, Ph.D. Thesis, University of Pennsylvania, May 1965].

Then

$$\left(\frac{\partial \ln i}{\partial 1/T}\right) = -\left(\frac{\Delta H_{\eta=0}^{\ddagger}}{R}\right) + \left(\frac{\alpha \eta F}{R}\right) \tag{148}$$

since $(\partial E^0/\partial T)$ is negligible and α is assumed independent of T. Thus,

$$\Delta H_{\eta=0}^{\ddagger} = \Delta H_{\eta}^{\ddagger} + \alpha \eta F \tag{149}$$

The values of $\Delta H_{\eta=0}^{\ddagger}$ were obtained by this theoretical extrapolation from an average $\Delta H_{\eta}^{\ddagger}$ at an average potential.

(*f*) *Other kinetic parameters*—Current–potential relations were determined galvanostatically and potentiostatically in solutions of different pH for various hydrocarbons to give the pH dependence of the oxidation reactions at Pt anodes. A typical example is seen in Figure 24 for the oxidation of ethylene. A similar procedure was carried out at different partial pressures to obtain the reaction

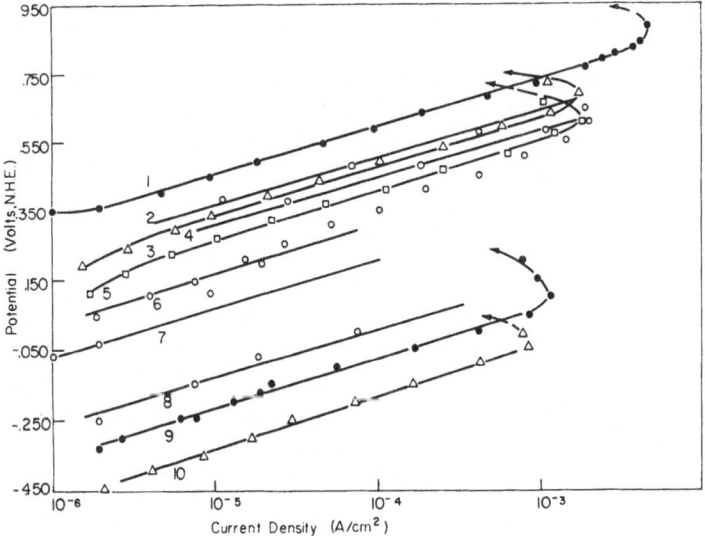

Figure 24. Current potential relations as a function of pH for ethylene oxidation at platinized platinum, 80°C: pH values: (1) 0.5, (2) 1.9, (3) 2.3, (4) 2.9, (5) 3.45, (6) 5.6, (7) 6.5, (8) 9.8, (9) 10.8, (10) 12.5 [from Wroblowa, Piersma, and Bockris, *J. Electroanal. Chem.* **6** (1963) 401].

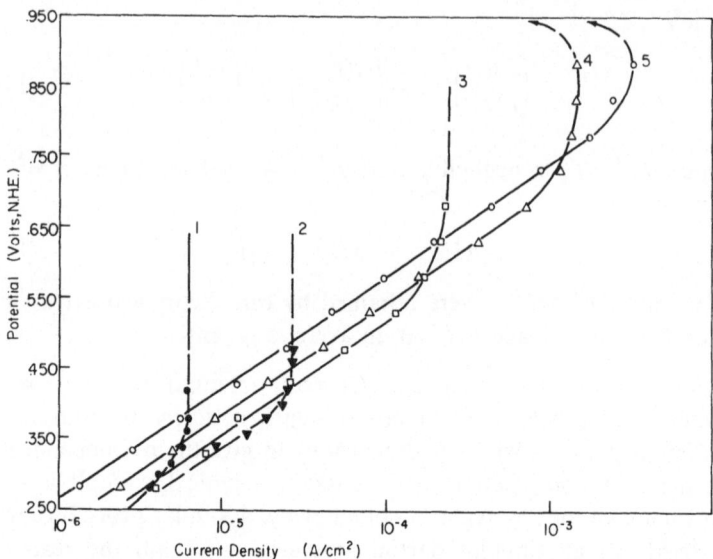

Figure 25. Current–potential relations as a function of partial pressure for ethylene oxidation at platinized platinum, $1\,N$ H_2SO_4, $80°C$: (1) 10^{-4} atm, (2) 10^{-3} atm, (3) 10^{-2} atm, (4) 10^{-1} atm, (5) 1 atm [from Wroblowa, Piersma, and Bockris, *J. Electroanal. Chem.* **6** (1963) 401].

order of the hydrocarbon (see, e.g., Figure 25). The kinetic parameters thus determined are collected in Table 17. The similarity in electrochemical behavior of these compounds can be expressed by the empirical relation

$$i = k\,C_{org}^{\gamma}\,C_{H^+}^{-0.5}\,\exp\!\left(\frac{FV}{RT}\right) \tag{150}$$

where $0.1 < \gamma < 0.25$, and k varies only about one order of magnitude from the fastest oxidation reaction (allene and ethylene) to the slowest (benzene and butadiene). These similarities indicate that (*a*) the same step is rate-controlling for the anodic oxidation of the unsaturated hydrocarbons discussed here, and (*b*) the position of this step in the reaction sequence is the same.

The above applies only for dilute sulfuric acid or sodium hydroxide solutions and platinum anodes. In concentrated phosphoric acid a positive reaction order with respect to the hydrocarbons has been observed.[68,262] On other electrodes, e.g., Au and Pd, a different Tafel slope has been reported[70] indicating a different mechanism. These cases will not be discussed further since they have not been fully investigated.

(g) *Mechanistic conclusions*—A number of possible reactions for the anodic oxidation of unsaturated hydrocarbons at platinum electrodes in dilute H_2SO_4 or NaOH have been considered, by Bockris and co-workers,[32,33,104] and the only rate-determining step which yields kinetic parameters that agree satisfactorily with the observed results is the water-discharge reaction. This may be represented, using the case of ethylene, as

$$C_2H_4 \rightleftharpoons C_2H_{4\,ad} \tag{151}$$

$$H_2O \xrightarrow{\text{r.d.s}} \cdot OH_{ad} + H^+ + e \tag{78}$$

$$C_2H_{4\,ad} + \cdot OH_{ad} \rightarrow C_2H_{3\,ad} + H_2O\,(\text{or}\,C_2H_4OH_{ad}) \tag{152}$$

$$C_2H_{3\,ad}\,(\text{or}\,C_2H_4OH_{ad}) + \cdots \rightarrow \cdots \rightarrow 2CO_2 + 11H^+ + 11e \tag{153}$$

The intermediate formed in the reaction of adsorbed ethylene and hydroxyl radical has not been defined. Since the reaction proceeds to CO_2 at about 100% efficiency, the intermediate species are probably strongly adsorbed free radicals which react rapidly in successive steps. The last equation given is meant to include all of the further reactions, including eleven electron transfer steps, required to yield the final products. The coverage by all intermediates is considered negligible in comparison with the coverage of the adsorbed reactant since they occur after the rate-limiting step and are not in equilibrium because of the constant removal of the final product, CO_2. The derived kinetic parameters for pH and pressure dependence will now be considered in greater detail.

(α) pH effect—As may be seen from the experimental results,[32,33,104] the anodic oxidation of unsaturated hydrocarbons occurs uniformly throughout the pH range from pH ≈ 0 to pH ≈ 12.5 (K water at 80°C is $10^{-12.5}$) although the change of solution pH in the region pH 5–8 limits the determination of current–potential curves in the region of intermediate pH. If the hydroxyl radical were to result from discharge of OH^-, then in $1 N H_2SO_4$, where pH ≈ 0, the concentration of OH^- would be $10^{-12.5}$ and a limiting current of the order of 10^{-14} A/cm^2 would be expected. Current densities of about 10^{-4} A/cm^2 (true area) were

Table 17

Kinetic Parameters for Oxidation of Some Unsaturated Hydrocarbons at Pt Anodes

Hydrocarbon	i_0 A/cm² (geom)	$\Delta H^{\ddagger}_{\eta=0}$ kcal/mole	Q_{CO_2}, %	b	$\left(\dfrac{d\log i}{d\log p}\right)_v$	$\left(\dfrac{d\log i}{d\,\mathrm{pH}}\right)_v$	$\left(\dfrac{dV}{d\,\mathrm{pH}}\right)_i$ mV
C_2H_4	5.0×10^{-8}	26.3	100	$2RT/F$	-0.2	0.45	-65
C_3H_4	2.1×10^{-8}	26.8	93	$2RT/F$	-0.17	0.46	-68
C_3H_6	2.0×10^{-8}	30.0	97	$2RT/F$	-0.14	0.46	-70
C_4H_6	0.5×10^{-8}	27.3	60–90	$2RT/F$	-0.13	0.39	-65
$C_4H_8 - 1$	1.2×10^{-8}	28.6	70	$2RT/F$	-0.16	0.47	-67
$C_4H_8 - 2$	1.2×10^{-8}	29.3	85	$2RT/F$	-0.20	0.47	-66
C_6H_6	0.2×10^{-8}	30.9	60–90	$2RT/F$	-0.11	0.40	-51

observed for ethylene oxidation. Thus, it must be concluded that the discharging species is the water molecule.

From the kinetic equation for the water-discharge mechanism

$$i = k(1 - \theta_T) \exp\left(\frac{\beta V F}{RT}\right) \tag{154}$$

it would appear that a pH dependence of zero is predicted from the simple model given. This fact does not militate against the mechanism but tends rather to lend it support because similar anomalous pH effects have been observed by other investigators who were concerned with simple reactions in which the charge transfer from H_2O (or OH^-) is certainly the rate-limiting step.[263-265] A summary of these results is given in Table 18.

A number of possibilities for the anomalous pH effect have been discussed,[32] and are summarized as follows:

1) Suppose that the water molecule undergoes charge transfer only when oriented with the oxygen atom towards the electrode, and that the potential of zero charge (p.z.c.) of Pt decreases as the pH increases. Then, the fraction of water molecules which could react increases with increasing pH at constant potential. This is in qualitative agreement with experimental observation. However, the changing orientation of water molecules would, under these conditions, give rise to a dipole potential contribution to the overall metal–solution potential difference and hence to deviations from the linear Tafel behavior, which is not observed experimentally.

Table 18

Experimental and Theoretical pH Dependence in Certain Reactions Involving the Discharge of H_2O or OH^- in the Presence of Excess Neutral Salt

Origin	$\left(\dfrac{d \log i}{d\,\mathrm{pH}}\right)_V$	$\left(\dfrac{d\eta}{d\,\mathrm{pH}}\right)$	$\left(\dfrac{d \log i_0}{d\,\mathrm{pH}}\right)$
Theoretical consideration {discharge from H_2O	0	RT/F	$-\alpha$
of simple model {discharge from OH^-	1	$-RT/F$	α
O_2 evolution on Pt in H_2SO_4 (Ref. 263)	0.25	$\alpha RT/F$	—
O_2 evolution on Pt over wide pH range (Ref. 264)	—	0	—
O_2 evolution on Pd and Au alloys (H_2SO_4, NaOH) (Ref. 265)	—	0	0
Ethylene oxidation (Ref. 32)	0.45	~ 0	~ 0

2) Assume that adsorbed OH results from the discharge of both H_2O and OH^-. Then,

$$i = (k_1 c_{H_2O} + k_2 c_{OH^-})(1 - \theta) \exp\left(\frac{\beta V F}{RT}\right) \tag{155}$$

and

$$\left(\frac{d \ln i}{d \, pH}\right)_V = \frac{1}{k_1 c_{H_2O}/(k_2 c_{OH^-}) + 1} \tag{156}$$

which is not in agreement with the observed pH dependence.

3) From several experimental observations, there is good indication that the potential of zero charge of Pt is dependent on pH.[140,172,266,267] Then, for a given value of the measured Galvani potential difference (total p.d., across the double layer), the Volta p.d. (charge dependent p.d.) changes, i.e., a pH dependence of the current densities at constant potential would be found. From experimental observation, an empirical relation for the pH dependence of the p.z.c. has been obtained:[140,172]

$$E_{p.z.c.} = E^0_{p.z.c.} + \frac{RT}{F} \ln a_{H^+} \tag{157}$$

When this is placed in the rate equation [equation (154)],

$$i = k(1 - \theta_T) \exp\left\{\left[V - \left(V^0_{p.z.c.} + \frac{RT}{F} \ln a_{H^+}\right)\right]\left(\frac{\beta F}{RT}\right)\right\} \tag{158}$$

and

$$\left(\frac{d \log i}{d \, pH}\right)_V = \beta \approx 0.5$$

which is the observed value. Thus, the apparently anomalous pH effect is well explained by the pH dependence of the potential of zero charge of Pt. However, the cause of this pH dependence of the p.z.c. has not been established, but the experimental observations appear to be valid.

(β) Pressure effect—Since the negative pressure dependence observed in the anodic oxidation of unsaturated hydrocarbons is a somewhat unexpected result, i.e., the reaction orders with respect to organic species are positive for the anodic oxidation of most organic species, including saturated hydrocarbons, the question of the validity of these observations arises. That the negative pressure effect is real and does not represent artifacts, e,g., the effect of the adsorption of traces of impurities or the formation of a poisoning species which may be related to the observed time effects, is shown by the following:

1) The pressure effects are reversible on repeated pressure changes without hysteresis.

2) Positive pressure effects have been reported for ethylene in systems identical except for the use of Au or Pd rather than Pt electrodes.[70] For these cases, a different mechanism has been proposed.

3) The adsorption of the poisoning impurities or the formation of a polymer which covers the electrode surface would lead in most cases to increased Tafel slopes.

4) Evidence that the negative pressure effect is probably not related to the time effects is obtained from the fact that similar time effects are observed in the anodic oxidation of organic species which have positive concentration dependence, e.g., HCOOH, $H_2C_2O_4$, or CH_3OH.

5) The negative pressure effect has been observed for a number of gaseous hydrocarbons (and for a liquid, benzene) which have widely different impurities. A negative pressure effect has been reported in an independent study of acetylene oxidation.[166] At high methanol concentrations, the reaction rate for methanol oxidation was found to decrease with increasing CH_3OH concentration.[268,269] A further examination of the pressure effect can be made from a consideration of the reaction order for the unsaturated hydrocarbons. For the adsorption of large organic species a Langmuir-type isotherm can be written, as a first approximation,

$$\frac{\theta}{(1 - \theta)^n} = Kp \qquad (159)$$

where n is the number of sites required for the adsorption of one organic molecule, i.e., $1/(1 - \theta)^n$ is the probability of finding n adjacent unoccupied sites. For the high coverages reported for ethylene adsorption ($\theta_{C_2H_4} > 0.9$), θ will be essentially invariant with ethylene partial pressure (over the limited range of pressures considered); however, $(1 - \theta)$ can change considerably. Thus, one has approximately

$$(1 - \theta) \alpha \, p^{-1/n} \tag{160}$$

and by substitution into the rate equation the reaction order is found to be

$$\left(\frac{d \log i}{d \log p}\right)_V = -\frac{1}{n} \tag{161}$$

Taking $n = 4$, as discussed previously for ethylene adsorption, the reaction order for ethylene is predicted to be -0.25 as compared with the experimental value of -0.20. From Table 17, variation of the reaction order of the organic species for the anodic oxidation can be observed. With reference to last equation, it is evident that different reaction orders are expected if the organic species require different numbers of adjacent, free, metal sites for adsorption. For example, butadiene, which has two conjugated double bonds, should require approximately twice as many sites as ethylene for adsorption. This is in fact observed since

$$\left(\frac{d \log i}{d \log p}\right)_V = -\frac{1}{n} = -0.13$$

giving $n = 8$ sites, as compared with $n = 4$ for ethylene. A comparison of the experimental and predicted values of n is given in Table 19. Within the accuracy of the experimental results and in

Table 19

Correlation of the Number of Metal Sites Required for Hydrocarbon Adsorption Obtained Experimentally from Pressure Dependence with Predicted Values

Hydrocarbon	n experimental	n predicted
C_2H_4	5	4
C_3H_4	6	6
C_3H_6	7	6
C_4H_6	8	8
$C_4H_8 - 1$	6	6
$C_4H_8 - 2$	5	8
C_6H_6	9	8

absence of further information on the structures of the adsorbed organic species, the correlation is satisfactory and lends further support to the kinetic treatment given.

(γ) Control in the "passivation region"—A comparison of the current potential curves for oxidation of the various hydrocarbons (Figure 22) shows that the potential of "passivation" is essentially constant with the exception of allene and butadiene which show passivation beginning about 0.1 V more anodic. It appears that allene and butadiene delay the formation of oxide, perhaps because of a more anodic desorption potential for these two compounds. A comparison of the capacity–potential curves for ethylene and butadiene[33] (Figure 26) shows an increase of the double-layer capacity in the potential region of passivation which is more anodic in the presence of butadiene. This increase in capacity corresponds to a pseudocapacity of the oxide formed in this region,

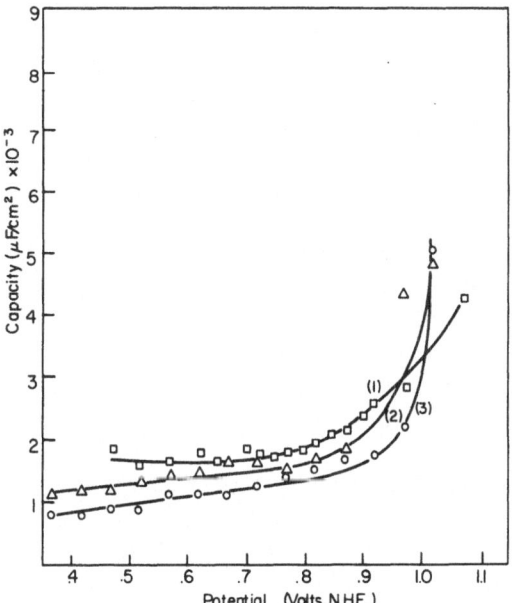

Figure 26. Capacity as a function of potential at platinum anodes in $1 N$ H_2SO_4: (1) butadiene, (2) nitrogen, (3) ethylene [from Bockris, Wroblowa, Gileadi, and Piersma, *Trans. Faraday Soc.* **61** (1965) 2531].

and the shift observed supports the view that butadiene and allene delay the formation of oxide.

Further support for the theory that the quasi-passivation region is related to oxide formation is found in the study of Dahms and Bockris.[70] From a study of ethylene oxidation on several noble metals, they observed that passivation occurred at different potentials, which, however, always coincided with potentials on the various metals at which a fraction of approximately 0.2 of the electrode surface was covered with oxygen.

(δ) Time effect—The time dependence of the current (in potentiostatic measurements) or potential (in galvanostatic measurements) appears to be a general phenomenon associated with electrode reactions. In a review of hydrogen overpotential, Hickling and Salt concluded that the variation of overpotential with time is a fundamental feature of overpotential phenomena at certain cathodes and is not due to inadequate experimental control.[270] Bockris also reported that his results for the increase of overpotential with time supported the contention that this time effect is a fundamental phenomenon of hydrogen overpotential.[271] Explanations advanced for the variation of hydrogen overpotential with time on platinum are, (a) progressive inactivation of active centers of the metal,[272,273] and (b) slow rate of attainment of equilibrium adsorption.[274]

Recent measurements of the diffusion of electrolytically generated hydrogen into platinum[275] and the associated change of overpotential with time lend support to the view that the time effects observed on Pt (and possibly on other metals which absorb hydrogen, e.g., Ni, Fe) are due to slow saturation of the bulk of the metal with atomic hydrogen, associated with possible phase formation.

The time variation of the oxygen overpotential at platinum has been examined by Busing and Kauzmann using the method of potential decay.[276] They concluded that most of the time variation could be understood if a nonuniform electrode surface with the electrode process occurring at relatively few active centers was assumed. The number of active centers was considered potential dependent and the time variation was attributed to poisoning of these active sites by solution impurity or by intermediates.

The time variation observed in anodic organic oxidations is much more pronounced than for the hydrogen or oxygen electrodes. Times of the order of 10 min were required to reach steady-state currents for formic acid oxidation.[39] Steady states required 30–60 min for oxalic acid oxidation[226] and for acetylene oxidation,[193] and several hours were required for ethylene oxidation in alkaline solution.[69] The results obtained by Bockris, Wroblowa, Gileadi, and Piersma show that times varying from a few minutes to several hours, depending on the potential region, are required to attain steady-state behavior for the anodic oxidation of unsaturated hydrocarbons.[32,33] Few attempts have been made to examine the time effects associated with organic oxidation (but cf. Ref. 69) and an acceptable explanation has only recently been advanced.[33] From correlation of a number of proposed mechanisms with the large amount of experimental evidence Bockris and co-workers have concluded[33] that the initial time effects (up to 15–20 min) are due to diffusion of the organic species to the electrode, followed, at larger times, by a slow filling up of the remaining free surface by activated adsorption.[104] The time dependence of the current for ethylene oxidation has been examined using equations derived by Delahay and Trachtenberg.[277] Thus the variation of surface concentration Γ_t with time may be written as

$$\Gamma_t = \Gamma_{eq}\left[1 - \exp\left(-\frac{Dt}{K\delta}\right)\right] \qquad (162)$$

where D is the diffusion coefficient of the adsorbing species in solution, K is the equilibrium constant for adsorption, and δ is the diffusion layer thickness.

For the water discharge mechanism, the rate equation has been given as

$$i = k(1 - \theta_T)\exp\left(\frac{\beta V F}{RT}\right) \qquad (154)$$

Combining with equation (162) and assuming $\theta_{C_2H_4} = \theta_T$, we obtain

$$i = k\left\{1 - \theta_{eq}\left[1 - \exp\left(-\frac{Dt}{K\delta}\right)\right]\right\}\exp\left(\frac{\beta V F}{RT}\right) \qquad (163)$$

At high values of the coverage where $\theta_{eq} \doteq 1$, this reduces to

$$i = k \exp\left(-\frac{Dt}{K\delta}\right) \exp\left(\frac{\beta V F}{RT}\right) \tag{164}$$

Differentiating equation (164) with respect to time at constant potential one has

$$\left(\frac{di}{dt}\right)_V = -\frac{kD}{K\delta} \exp\left(-\frac{Dt}{K\delta}\right)\left[\exp\left(\frac{\beta V F}{RT}\right)\right] \tag{165}$$

hence

$$\ln\left(-\frac{di}{dt}\right)_V = \ln\frac{kD}{K\delta} + \frac{\beta F V}{RT} - \frac{Dt}{K\delta} \tag{166}$$

and

$$\left[\frac{d\ln(-di/dt)_V}{dt}\right]_V = -\frac{D}{K\delta} \tag{167}$$

Plots of $\log(-di/dt)_V$ vs. t were found to be linear up to about 10 min.[33] The slopes obtained were essentially independent of potential and numerically in agreement with values calculated using values of the adsorption coefficient[94] K and of D determined independently.[170] Thus, the variation of current with time at constant potential is due to slow attainment of adsorption equilibrium under diffusion control. The break in the curves after about 10 min is indicative of a different cause for the time effects at longer times, which probably involves slow, activation-controlled adsorption on a small fraction of the electrode surface.

Fast chemisorption followed by slow activated adsorption involving only small amounts of the adsorbent have been reported from studies in the gas phase. Trapnell reports that adsorption may be nonactivated in its initial stages, but becomes activated when more than about half of the surface becomes covered.[181] Experimental evidence for mixed chemisorption (activated adsorption) and diffusion control has been reported by Ward for the adsorption of hydrogen on copper.[278]

(iii) Alkanes

Although a number of studies showing that saturated hydrocarbons can be oxidized electrochemically have appeared in the literature, no systematic mechanism studies have been reported.

Much of the work with these compounds has been carried out with fuel cells and the emphasis has been on power output and performance. Therefore, these studies with alkanes will not be further discussed here.

7. Other Organic Compounds

The anodic behavior of a number of other organic compounds has recently been reported, particularly the reversible electrochemical reactions of aromatic systems. Much of this work has been carried out using the methods of cyclic voltametry and chronopotentiometry as previously discussed. Adams and his co-workers have been active in this area, with studies of the anodic oxidation of p-methoxyphenol,[279] N, N-dimethylaniline,[280,281] N-methylaniline,[282] and N, N-dimethyl-p-toluidine,[282] for example. Trimethylamine,[160] phenols,[283-286] aromatic amines,[283,285,287-290] and a number of aromatic hydrocarbons[291] have been examined using cyclic voltametry or other transient methods, in several instances with a rotating anode. In most of these investigations kinetic parameters have not been clearly determined nor have reaction mechanisms been established. The polarographic oxidation of a number of compounds at solid electrodes, or simply cyclic voltametry, was reported by Pysh and Yang[292] and by Suantoni et al.[293]

The electrochemical processes involved in the quinone-hydroquinone redox couple have recently been examined by Bagotzky et al.[294] In acid solutions the rate determining step was found to be

$$Q_{ad} + H^+ + e \xrightarrow{\text{r.d.s.}} QH_{ad} \tag{167}$$

followed by

$$QH_{ad} + H^+ \rightleftharpoons QH_{2\,ad}^+ \tag{168}$$

and

$$QH_{2\,ad}^+ + e \rightleftharpoons QH_2 \tag{169}$$

In alkaline solution a similar mechanism was proposed, namely,

$$Q_{ad} + e \xrightarrow{\text{r.d.s.}} Q_{ad}^- \tag{170}$$

$$Q_{ad}^- + H^+ \rightleftharpoons QH_{ad} \qquad (171)$$

$$QH_{ad} + H_2O + e \rightleftharpoons QH_2 + OH^- \qquad (172)$$

V. THE INFLUENCE OF MOLECULAR STRUCTURE ON REACTION RATE

A discussion of the nature of electronic interactions between the metal surface and adsorbed molecules requires the consideration of several factors: the magnitude of the electronic work function, the number of unfilled electron levels in the 3d, 4d and 5d bands, the structure of adsorbing crystal planes, the electron affinity of the adsorbing species, the asymmetry of the electron configuration of the adsorbed species, the presence of unpaired electrons and π electrons, the effects due to steric hindrance, and the various bond energies in the adsorbing species. The first three are characteristics of the metal surface. The remaining factors must be considered in the interpretation of the influence of structure of the reactant organic species on the reaction rate. These will each be only briefly considered since there have been very few electrochemical studies designed to examine the effect of molecular structure.

1. Electrocatalysis

(i) A Rational Potential for Comparison of Catalytic Activity

Electrocatalysis is similar to heterogeneous catalysis for which considerable work has been carried out in the gas phase. The primary difference is the effect of potential in electrode reactions which can change the rates of intermediate steps including charge transfer and the energy of adsorption and thus the concentration of reactants, intermediates, and products adsorbed on the surface. An additional effect may be that of the solvent in electrochemical reactions which can alter the rate of reaction and may take part in the reaction (cf. p. 75). Since a given overall reaction rate may not vary with potential in the same way on different electrodes, i.e., different Tafel slopes may be obtained with different metals, it is necessary to decide upon some rational potential at which the catalytic powers of different metals may be compared. The most reasonable point of comparison is that at which the influence of the applied electric field is minimized, i.e., the potential at which the charge on the metal is zero. This potential of zero charge (p.z.c.)

was recognized early by the Russian School to be of fundamental importance both for the thermodynamics of adsorption and the kinetics of electrochemical reactions.[295] The importance of the potential of zero charge as a rational potential for the comparison of reaction rates has been discussed in detail by Antropov.[296] Dahms and Bockris have recently used the potential of zero charge to compare the catalytic activity of several metals for the electro-oxidation of ethylene. A similar approach has previously been adopted by Conway *et al.* for the interpretation of the catalytic activity of Cu, Ni, and a number of their alloys for the hydrogen evolution reaction.[265]

The rate of an electrochemical reaction may be expressed (in units of current density) by the general equation

$$i = i_0 \exp\left(\frac{\alpha \eta F}{RT}\right) \tag{173}$$

where η is the overpotential, and α is the transfer coefficient.* The exchange current density may be written as

$$i_0 = k_0 \exp\left(\frac{\alpha V_r F}{RT}\right) \tag{174}$$

The quantity V_r in equation (174) is the absolute metal-solution potential difference at the reversible potential, which cannot be determined experimentally. It is seen that the observed exchange current density (which is a practical measure of the catalytic activity) is due to a factor k_0, which is mainly chemical in nature, and an electrical term. It would therefore seem desirable to compare the reactivities of metals at their respective absolute null point, i.e., at an overpotential such that $V = 0$ for each metal, in order to understand the factors which determine the chemical specific rate constant k_0. Since V cannot in principle be measured, the best approximation is to compare electrochemical rates at the potential

---

*The transfer coefficient α is distinctly different from the symmetry factor β which has been used in some of the previous equations. The former is an experimental parameter derived from the observed i–V relationship while the latter is a quantity of fundamental importance, the magnitude of which depends on the shape of the potential energy barrier for the rate determining step in the reaction sequence. In general β has a value close to 0.5 while $\alpha \geqslant \beta$, and it may reach high values of 3–4 for certain mechanisms.

of zero charge. While the absolute metal–solution potential difference V is clearly not equal to zero at the p.z.c., it may be considered nearly the same for all metals, since the orientation of the solvent dipoles on the surface depends primarily on the charge on the metal and to a much lesser extent on specific interactions between the metal and the solvent.

The potential of maximum adsorption may be used as an alternative point of reference for comparison of the specific rate constants of a reaction on different metals. Thus it has been shown above (cf. p. 80) that at this point, equal numbers of water molecules are aligned in the two possible orientations and the net dipole potential contribution due to the aqueous solvent at the interface tends to zero. Theoretical calculations[141] show that the potential of maximum adsorption should correspond approximately to the same charge on all metals. Hence the absolute potential drop across the interface is nearly the same for all metals at their respective potential of maximum adsorption, and k_0, the chemical part of the electrochemical rate constant, can be conveniently compared using this potential as the point of reference.

The experimental methods for determination of the potential of zero charge have been reviewed by Swinkels,[299] Conway,[300] and by Argade and Gileadi.[266] Methods of measurement of electrosorption of organic compounds on solid electrodes have been discussed recently by Gileadi.[110]

(ii) Correlations from Gas Phase Heterogeneous Catalysis

Metals catalyze processes by virtue of their ability to adsorb reactants or products in an appropriate manner and provide a path with a lower energy of activation. It has long been appreciated that the activity of a heterogeneous catalyst chiefly resides at the interface between the metal and the less dense phase, and that the reactant (or at least one of the reactants if there are more than one) must be adsorbed at this interface. If the reactant is too strongly adsorbed, it will be correspondingly difficult to remove, and may thus result in reduced rates of reaction. If it is too weakly adsorbed, it will have little chance of remaining on the surface long enough to react.

An early interpretation of the function of the catalyst was that of Balandin who supposed that a catalytic reaction is determined by the power of attraction of the various parts of the reactant molecule for different catalyst atoms.[301] These short range forces were thus determined solely by the relative spacing of surface and reactant atoms. The role of the catalyst was then to assist breaking and making of bonds involved in reaction through orienting and deforming the reactants in the adsorbed layer, i.e., the activation energy should be a minimum when the energy of the transition state is a minimum, for which an optimum metal–metal bond distance will exist. This "Multiplet theory" of Balandin is incomplete, however, because it does not explain the complex features of catalytic specificity and poisoning and because it only considers physical adsorption. The energy of the transition state is also a function of the bond strengths between the various species and the metal atoms, i.e., their heat of adsorption, which will depend on electronic contributions of the metal.

A study of the reaction between hydrogen and ethylene on several faces of single crystals of nickel showed that the catalytic activity varied significantly with crystal face.[302] From these studies three factors were indicated as important: the spatial arrangement of surface atoms, electronic properties of the surface, and the nature and number of imperfections on the surface. Correlations of heats of adsorption, particularly for hydrogen, with the d-character (or alternatively with the number of d-electron vacancies) of the metal has achieved limited success. A "volcano" relation between the latent heat of sublimation and the catalytic activity of metals has been obtained in several reactions, e.g., formic acid decomposition, hydrogenation of hydrocarbons, ammonia exchange, and acetone hydrogenation. Other studies have indicated that cracks and fissures of the metal surface which gave it pronounced heterogeneous character for physical adsorption do not influence chemisorption bonds very much.[303]

For a meaningful discussion of electronic factors in catalysis it is necessary to briefly review the nature of chemisorption bonds.[182] Two theories of the metallic state have been accepted, the electron band theory and the valence bond theory. Both theories recognize the existence of two separate functions for valence electrons in metals; one function is to bind the atoms together and the other is to account for magnetic and conductive properties. In the electron band theory, as particularly applied to the transition metals, the s-electron energy band is broad with a low maximum electron level density (number of electrons having an energy in the range $E + dE$), whereas the d-electron band is narrow with a much higher maximum electron level density. The binding of metal atoms is attributed to overlapping d- and s-bands. Magnetic and conductive properties are associated with unoccupied states in the d-band (holes in the d-band), e.g., the number of unoccupied d-band states is numerically equal to the saturation moment in Bohr magnetons. The valence bond theory of Pauling assumes that s-, p-, and d-electrons are capable of taking part in cohesive bonding and that the actual bonds are d-s-p-hydridized orbitals formed by overlapping of the electron orbitals. The extent to which d-electrons participate in the d-s-p-binding orbital is the percentage d-character of the metallic bond. Magnetic and conductive properties are assigned to the electrons in atomic d-orbitals in the valence bond theory of transition metals. Although the electron band theory is a more fundamental approach, the valence bond theory being of a somewhat more empirical nature, the latter is generally preferred in a discussion of the behavior of transition metals.[182]

Two opposing theories on the nature of chemisorption bonds have been supported in recent years. Trapnell has subscribed to the view that the chemisorption bond is a covalence between electrons from the adsorbate and unpaired electrons in atomic d-orbitals.[181] This concept easily interprets the high chemisorption activity of the transition metals and the decrease of magnetic susceptibility which can follow chemisorption. The alternative view, supported by Dowden, is that the metal–adsorbate bond is essentially similar to the metal–metal bond, i.e., a free d-s-p-orbital is employed.[304] In this theory, the role of unpaired d-electrons is that of forming an intermediate without which the final state cannot be attained, unless a high activation energy is overcome or the molecule is previously dissociated to atoms.

The theory that the chemisorption bond is a covalence involving partially filled d-orbitals is not subject to quantitative test by calculation. Experimental correlations of heats of adsorption with the percentage d-character of the metal, particularly for adsorption of hydrogen on the transition metals, have led to taking the percentage d-character as a measure of unavailability of electrons in atomic d-orbitals and thus the expected strength of the chemisorption bond. For the alternative view in which the surface bond is supposed to be similar to a metal–metal bond, the bond energy may be readily calculated, for example, by use of the Pauling equation, if the latent heat of sublimation of the metal and the electronegativities are known. Thus, the experimental and calculated metal–hydrogen bond strengths have been calculated with the latent heat of sublimation of the metal.[182]

In a comparative assessment of these two views of chemisorption bonds, Bond has concluded that neither of the models, i.e., correlation of heats of adsorption with the degree of occupation of atomic d-orbitals as indicated by percentage d-character, nor correlation of heats of adsorption with the binding strength of d-s-p-hybridized orbitals, as given by the latent heat of sublimation, is completely successful.[182] For a given process, e.g., the adsorption of hydrogen, no single physical property of the bulk metal can be successfully correlated. With all the quantitative information on chemisorption, however, this is not altogether surprising since correspondence between metal atoms in the bulk and on the surface may not be complete.

(iii) Empirical Characteristic of Electrocatalysis

Müller made one of the first attempts to study the role of the metal in the electrochemical oxidation of organic compounds in his investigations of formic acid and methanol oxidations.[23] Other studies of the electrocatalysis for organic reactions are few. A comparison of the catalytic activity of platinum black and noble metal alloys for the anodic oxidation of methanol has been made by measuring the potentials established at these electrodes at a constant current density of $20 \, mA/cm^2$.[305] Breiter has examined the catalytic activity of noble metals for methanol oxidation as indicated by relative current peak heights of the first oxidation wave during the anodic sweep using cyclic voltametry.[306] Dahms and Bockris have made the only attempt to determine catalytic activity for an organic electrode reaction at a rational potential, i.e., at the p.z.c.[70] They found that the metals studied fall into two groups of electrocatalytic activity: Pt, Rh, and Ir, which oxidized ethylene quantitatively to CO_2, and Au and Pd, which gave incomplete oxidation and also different kinetic parameters. The difference in behavior of the two groups was explained by a difference in metal organic bond strengths which may be related to the distinct difference of heats of sublimation for the metals of the two groups. The different catalytic activities within the groups were correlated with the differences in work function, the metal with the higher work function exhibiting the faster reaction.

Further studies in electrocatalysis are limited mostly to the hydrogen evolution reaction. Conway and Bockris found correlations between the overpotentials of metals at a constant current density of 10^{-3} A/cm^2, and the metal–hydrogen bond strengths, calculated for most of the metals.[307] For the transition metals a relation was found to exist between the exchange current densities and the percentage d-character of the metals, with i_0 increasing with increasing d-character.* A similar relation was found between the exchange current densities and the electronic work function of the transition metals. In a latter paper Bockris and Wroblowa[308] have discussed the catalytic activities for various metals and showed that the rates of reactions should preferably be compared at the potential of zero charge for each metal. Conway, Beatty, and DeMaine have extended the correlation of rate with d-character by using a series of Ni–Cu alloys to change the d-character from 0 for Cu to 0.6 for Ni.[265] A plot of the exchange current densities vs. the number of d-holes per atom gave a linear relation with log i_0 increasing with increasing number of d-holes. Similar correlation between exchange currents and d-character of the metal electrodes has also been observed from studies of the oxygen evolution reaction.

(iv) Summary

(a) Geometric factors—For the metals usually studied in the electro-oxidation of organic compounds, Pt, Pd, Rh, Ir, and Au, the lattice parameters are quite similar. These metals all have face-centered cubic structures with atomic radii within 0.1 Å. The effect of grain size as indicated by platinization of platinum does not appear to affect the catalytic activity. In a study of formic acid oxidation, Gottlieb found that the only effect of platinization was to increase the true area of the electrode and that there was no introduction of

---

*The decrease in reaction rate with increasing bond strength is interpreted in this case to indicate the electrochemical mechanism. Thus for a desorption step rate determining, the stronger the bond strength, the slower will be the rate of desorption and the rate of the overall reaction.

catalytic sites specific for the oxidation reaction.[210] Results obtained for hydrocarbon oxidations also indicate that platinization only increases the active area of the electrode without changing the catalytic activity.[104] This indicates that increased roughness of the electrode surface does not affect the catalytic activity as is observed in gas phase catalysis when chemisorption is involved. The effect of defect concentration in electrocatalysis has not been examined for organic oxidation. One study of the hydrogen evolution reaction on iron indicates that if 0.1 % carbon is introduced into zone-refined iron, the rate constant is increased by a factor of ten.[309] In general, whereas geometric factors can become important if the internuclear distances of the metal atoms are not within a certain range, particularly if more than one site attachment is involved in adsorption,[310] they play a small role and in many cases cannot be separated from the more important electronic factors.

(b) *Electronic factors*—The principal electronic factors associated with active metal catalysts, particularly the transition metals, are the electronic work function and the number of unfilled electron levels in the *d*-bands. The work function, which is a measure of the ease with which electrons are removed from the metal surface, appears to be less important than the number of *d*-vacancies or the binding strength of hybrid orbitals as indicated by the latent heat of sublimation. These factors are effective in determining the heats of adsorption of the organic species on the electrode, which decreases with decreasing number of *d*-electron vacancies in the metal surface or with increasing work function and decreases with decreasing heat of sublimation of the metal. While correlations between catalytic activity and electronic structure are few, some success has been obtained as previously indicated. Initial results for electrocatalytic studies of ethylene oxidation indicate that a "volcano" type of relation may exist between activity and the latent heat of sublimation of the metals studied.[311]

(c) *Other factors which may increase catalytic activity*—The primary aim of electrocatalysis is to increase the rates of electrochemical reactions to the point where mass transport becomes rate controlling while the overpotentials remain negligible. The obvious procedure is to examine *rationally chosen* electrode materials to find the most active for the reaction in question. Of the metals investigated for the anodic oxidation of organic compounds, platinum or possibly alloys of platinum with rhenium and ruthenium have exhibited the highest catalytic activity. The possibilities of alloys have been only little investigated. Some attempts have been made in the use of unconventional catalysts, e.g., nickel boride electrodes are used with good success in the hydrogen–oxygen fuel cell, and for hydrazine oxidation. Chelates have been reported catalytically active for hydrocarbon oxidation. By a suitable choice of metal ion and ligand, a series of chelates of varying coordinating tendency may be prepared. In addition to nickel boride, a large field of semiconductors can be examined. Control of the electron energy levels in a semiconductor by the choice of the suitable donor and acceptor species make possible a particularly well-controlled model system for electrocatalysis studies.

The decrease of electrode activity with time observed in most organic oxidations is another problem for the development of fuel cells. Studies which have been reported[202] indicate that electrodes can be activated by the application of suitable anodic pulses to the electrode. Activation by radioactive materials is being examined and shows promise, particularly for oxygen electrodes. Oxygen electrodes activated with radioactive Tl exhibited less polarization than inactivated electrodes. The use of photo- and sono-effects in this respect has also been suggested.

REFERENCES

[1] Moos, *Fuel Cells*, Ed. Young, Vol. 2, p. 3, Reinhold Publishing Company, New York, 1963.
[2] Heath and Worsham, *ibid.* p. 182.
[3] Bockris and Conway, *Record Chem. Progr.* **25** (1964) 31.
[4] Law and Perkin, *Trans. Faraday Soc.* **1** (1905) 31.
[5] *Gmelins Handbook*, Vol. 1 (1848).
[6] Allen, *Organic Electrode Processes*, Reinhold Publishing Company, New York, 1958.
[7] Faraday, *Pogg. Ann.* **33** (1834) 438.
[8] Kolbe, *Ann. Chem.* **69** (1849) 257.
[9] Fichter, *Trans. Electrochem. Soc.* **45** (1924) 107.
[10] Friedel, *Ann. Chem.* **112** (1859) 376.
[11] Bradt and Opp, *Trans. Electrochem. Soc.* **59** (1931) 237.
[12] Renard, *Compt. Rend.* **91** (1880) 125.
[13] Gotterman and Friedrichs, *Chem. Ber.* **2** (1897) 1942.

[14] Drechsel, J. Prakt. Chem. **29** (1884) 229; **31** (1886) 135; **38** (1888) 65.

[15] Nathansohn, Kolloidchem. Beih. **11** (1919) 261.

[16] Fichter, Z. Elektrochem. **27** (1921) 487.

[17] Haber, ibid. **4** (1898) 506.

[18] Hickling, Trans. Faraday Soc. **38** (1942) 27.

[19] Fichter, Trans. Electrochem. Soc. **45** (1924) 107.

[20] Fichter, ibid. **56** (1929) 8.

[21] Schall, Z. Elektrochem. **3** (1896) 83.

[22] Fichter, Trans. Electrochem. Soc. **76** (1939) 309.

[23] Müller, Z. Elektrochem. **28** (1922) 101; **29** (1923) 264.

[24] Müller and Hindemith, ibid. **33** (1927) 561.

[25] Müller and Schwab, ibid. **33** (1927) 568.

[26] Müller and Tanaka, ibid. **34** (1928) 256; **35** (1929) 38.

[27] Müller and Schwabe, ibid. **34** (1928) 170.

[28] Müller and Takegume, ibid. **34** (1928) 704.

[29] Glasstone and Hickling, J. Chem. Soc. (1934) 1878; (1936) 820.

[30] Glasstone and Hickling, Trans. Electrochem. Soc. **72** (1939) 333.

[31] Glasstone and Hickling, Chem. Rev. **25** (1939) 407.

[32] Wroblowa, Piersma, and Bockris, J. Electroanal. Chem. **6** (1963) 401.

[33] Bockris, Wroblowa, Gileadi, and Piersma, Trans. Faraday Soc. **61** (1965) 2531.

[34] Glasstone and Hickling, Electrolytic Oxidation and Reduction: Inorganic and Organic, Chapman and Hall Ltd., London, 1935.

[35] Fichter, Organische Elektrochemie, T. Steinkoff, Dresden & Leipzig, 1942.

[36] Brockman, Electro-organic Chemistry, John Wiley and Sons, New York, 1926.

[37] Conway and Dzieciuch, Can. J. Chem. **41** (1963) 21, 38, 55.

[38] Dickinson and Wynne-Jones, Trans. Faraday Soc. **58** (1962) 382, 388, 400.

[39] Brummer and Makrides, J. Phys. Chem. **68** (1964) 1448.

[40] Crum-Brown and Walker, Ann. Chem. **261** (1891) 107.

[41] Clarke, Myers, and Acree, J. Phys. Chem. **20** (1916) 286.

[42] Kalinin and Stender, J. Appl. Chem. USSR **19** (1946) 1045.

[43] Standard Oil Co., American Patent 1875310.

[44] Bhattachayya, Muthana, and Patankai, J. Sci. Ind. Res. (India) **11B** (1953) 369.

[45] Hultman and Powell Corp., American Patent 1992309.

[46] British Celanese Ltd., Bristish Patent 609, 594.

[47] Athanasiu, Bull. Chim. Para. Aplic. **31** (1926) 75.

[48] Athanasiu, Ber. Itsch. Chem. Ges. **64** (1931) 252.

[49] Nabuco de Arauja, Jr., Chem. Zbl. I (1934) 3394; II (1934) 2748.

[50] Athanasiu, Natl. Petrol. News **25** (1933) 22, 24, 26.

[51] Nabuco de Araujo, Jr., Chemica (Brazil) **1** (1933) 281.

[52] Schlatter, American Chemical Society National Meeting, Chicago, 1961; New York, 1963.

[53] Schlatter, Fuel Cells, Ed. Young, Vol. 2, p. 190, Reinhold Publishing Company, New York, 1963.

[54] Grubb, Proc. 16th Annual Power Sources Conf. p. 31, 1962.

[55] Niedrach, J. Electrochem. Soc. **109** (1962) 1092.

[56] Grubb and Niedrach, Proc. 17th Annual Power Sources Conf. p. 69, 1963.

[57] Young and Rozelle, Fuel Cells, Ed. Young, Vol. 2, p. 216, Reinhold Publishing Company, New York, 1963.

[58] Buck, Griffith, MacDonald, and Schlatter, Proc. 15th Annual Power Sources Conf. p. 16, 1961.

[59] Grubb and Michalski, Electrochemical Society National Meeting, New York, 1963.

[60] Oswin, Hartner, and Malaspina, *Nature* **200** (1963) 256.
[61] Vaucher and Bloch, *Compt. Rend.* **254** (1962) 3676.
[62] Cairns and Bartosik, Electrochemical Society National Meeting, Toronto, 1964.
[63] Binder, Köhling, Krupp, Richter, and Sandstede, American Chemical Society National Meeting, New York, 1963; Electrochemical Society National Meeting, Toronto, 1964.
[64] Griffith and Rhodes, *Fuel Cells*, p. 32, American Institute of Chemical Engineers, New York, 1963.
[65] Grubb and Michalske, *Nature* **201** (1964) 287.
[66] Grubb and Michalske, Proc. 18th Annual Power Sources Conf. 1964.
[67] Binder, Köhling, Krupp, Richter, and Sandstede, *J. Electrochem. Soc.* **112** (1965) 356; *Electrochim. Acta* **8** (1963) 781.
[68] Savits and Frysinger, Electrochemical Society National Meeting, Washington, 1964.
[69] Green, Weber, and Drazic, *J. Electrochem. Soc.* **111** (1964) 721.
[70] Dahms and Bockris, *ibid.* **111** (1964) 728.
[71] Schreiner, *Z. Elektrochem.* **36** (1930) 953.
[72] Kappanna and Jashi, *J. Indian Chem. Soc.* **29** (1952) 69.
[73] Karabinos, *Euclides* **14** (1954) 211.
[74] Choudhury and Khundkar, *Pakistan J. Sci. Res.* **4** (1952) 103.
[75] Choudhury and Khundkar, *Pakistan J. Sci. Res.* **5** (1953) 77.
[76] Bockris, Piersma, and Gileadi, *Electrochim. Acta.* **9** (1964) 1329.
[77] Pavela, *Ann. Acad. Sci. Fennicae, Ser. A, II, Chem.*, No. 59 (1954) 1.
[78] Reinmuth, *Anal. Chem.* **32** (1960) 1514.
[79] Nicholson and Shain, *Anal. Chem.* **36** (1964) 706.
[80] Erschler, *J. Phys. Chem. (USSR)* **22** (1948) 683.
[81] Gerischer, *Z. Elektrochem.* **54** (1950) 362; **55** (1951) 98.
[82] Vetter, *Z. Physik. Chem.* **194** (1950) 284; *Z. Elektrochem.* **55** (1951) 121.
[83] Vetter, *Elektrochemische Kinetik*, Springer Verlag, Berlin, 1961.
[84] Parsons, *Trans. Faraday Soc.* **47** (1951) 1332.
[85] Bockris, *Modern Aspects of Electrochemistry*, Ed. Bockris and Conway, Vol. I, Chap. 4, Academic Press Inc., New York, 1954.
[86] Delahay, *New Instrumental Methods of Electrochemistry*, Interscience Publishers, Inc., New York, 1954.
[87] Lingane, *Electroanalytical Chemistry*, Interscience Publishers, Inc., New York, 1958.
[88] Kolthoff and Lingane, *Polarography*, 2nd ed., Interscience Publishers Inc., New York, 1952.
[89] Gileadi and Srinivasan, *J. Electroanal. Chem.* **7** (1964) 452.
[90] Gileadi and Conway, *Modern Aspects of Electrochemistry*, Ed. Bockris and Conway, Vol. 3, Butterworths, Washington, 1964.
[91] Frumkin, *Z. Physik* **35** (1926) 792.
[92] Devanathan, *Proc. Roy. Soc. London, Ser. A* **264** (1961) 133.
[93] Frumkin and Damaskin, *Modern Aspects of Electrochemistry*, Ed. Bockris and Conway, Vol. 3, Butterworths, Washington, 1964.
[94] Gileadi, Rubin, and Bockris, *J. Phys. Chem.* **69** (1965) 3335.
[95] *Basic Studies of Sorption of Organic Fuels During Oxidation at Electrodes*, Final Report M64–341 prepared by American Oil Research and Development for U.S. Army Research Office, Durham, N.C.
[96] Shlygin and Frumkin, *Acta Physicochim. USSR* **3** (1935) 791.
[97] Brodd and Hackermann, *J. Electrochem. Soc.* **104** (1957) 704.
[98] Grahame, *Chem. Rev.* **41** (1947) 441.

[99] Breiter, Knorr, and Volkl, *Z. Electrochem.* **59** (1955) 681.
[100] Will and Knorr, *Z. Electrochem.* **64** (1960) 258.
[101] Schuldiner and Warner, *J. Electrochem. Soc.* **112** (1965) 212.
[102] Breiter and Gilman, *J. Electrochem. Soc.* **109** (1962) 622.
[103] Hickling and Wilson, *Nature* **164** (1949) 673.
[104] Piersma, Ph.D. Thesis, University of Pennsylvania (1965).
[105] Breiter, *Electrochim. Acta* **7** (1962) 533.
[106] Greene and Leonard, *Electrochim. Acta* **9** (1964) 45.
[107] Oikawa and Mukaibo, *J. Electrochem. Soc. Japan* **20** (1952) 568.
[108] Frumkin and Sothern, *J. Phys. Chem.* **58** (1954) 951.
[109] Schuldiner and Warner, *Electrochim. Acta* **11** (1966) 307.
[110] Gileadi, *J. Electroanal. Chem.* **11** (1966) 137.
[111] Laitinen, *Anal. Chem.* **33** (1964) 1458.
[112] Bockris, Devanathan, and Müller, *Proc. Roy. Soc. London, Ser. A* **274** (1963) 55.
[113] Müller and Nekrasow, *Electrochim. Acta* **9** (1964) 1015.
[114] Levich, *Physicochemical Hydrodynamics*, p. 335, translation from Russian by Scripta Technica, Inc., Prentice Hall, Englewood Cliffs, N.J. 1962.
[115] Balaschova and Merkulova, Moscow Conference on Electrochemistry, 1956.
[116] Balaschova, *Z. Physik. Chem.* **207** (1957) 340; *Zh. Fiz. Khim.*, **32** (1958) 22, 2266.
[117] Balaschova, Iwanow, and Kasarinow, *Dokl. Akad. Nauk. SSSR* **115** (1957) 336.
[118] Kolotyrkin and Medwedewa, *Zh. Fiz. Khim.* **31** (1957) 2669.
[119] Hevesy and Wiess, *Arg. Akad. Wiss.* **124** (1915) 131.
[120] Erbacher, *Z. Phys. Chem.* **A163** (1933) 196, 215; *Z. Elektrochem.* **44** (1938) 594.
[121] Hackermann and Stephens, *J. Phys. Chem.* **58** (1954) 904.
[122] Joliot, *Compt. Rend.* **184** (1927) 1325.
[123] Hutchinson, *J. Colloid Sci.* **4** (1949) 600.
[124] Dixon, Weith, Argyle and Sally, *Nature* **163** (1949) 845.
[125] Aniansson and Lanm, *Nature* **165** (1950) 357.
[126] Aniansson, *J. Phys. Chem.* **55** (1951) 1286.
[127] Cook, *Rev. Sci. Instr.* **27** (1950) 1081.
[128] Cook and Ries, *J. Phys. Chem.* **63** (1959) 226.
[129] Kafalas and Gatos, *Rev. Sci. Instr.* **29** (1958) 47.
[130] Schwabe, *Chem. Tech.* **10** (1958) 469.
[131] Weismantle, Ph.D. Thesis, Dresden, Germany (1958); *Keminergie* **2** (1959) 909.
[132] Blomgren and Bockris, *Nature* **186** (1960) 305.
[133] Schwabe, *Electrochim. Acta* **3** (1960) 186; **6** (1962) 223.
[134] Weissmantle, Schwabe, and Hecht, *Werket u. Knos* **12** (1961) 353.
[135] Schwabe and Schwenke, *Electrochim. Acta* **9** (1964) 1003.
[136] Dahms, Green, and Weber, *Nature* **196** (1962) 1310.
[137] Wroblowa and Green, *Electrochim. Acta* **8** (1963) 679.
[138] Dahms and Green, *J. Electrochem. Soc.* **110** (1963) 1075.
[139] Green, Swinkels, and Bockris, *Rev. Sci. Instr.* **33** (1962) 18.
[140] Bockris, Green, and Swinkels, *J. Electrochem. Soc.* **111** (1964) 743.
[141] Bockris and Swinkels, *ibid.* **111** (1964) 736.
[142] Conway, Barradas, and Zawidsky, *J. Phys. Chem.* **62** (1958) 676.
[143] American Oil Company, Whiting Laboratories, Quarterly Reports. 1–9, contract No. DA-11-022-ORD-4023.
[144] Reddy and Bockris, *Nat. Bur. Std. U.S. Misc. Publ.* **256** (1964) 229.
[145] Bockris, Devanathan, and Reddy, *Proc. Roy. Soc. London, Ser. A* **279** (1964) 327.
[146] Bockris, Genshaw, and Reddy, in press.

172 Electrochemical Oxidation of Organic Fuels

[147] Horiuti and Ikusima, *Proc. Imp. Acad. Tokyo* **15** (1939) 39.
[148] Horiuti, *J. Res. Inst. Catalysis* **1** (1948) 8.
[149] Vetter, Z. *Elektrochem.* **59** (1955) 596.
[150] Delahay, *J. Chim. Phys.* **54** (1947) 369.
[151] Delahay, *Advances in Electrochemistry and Electrochemical Engineering*, Ed. Delahay and Tobias, Vol. 1, Interscience Publishers, Inc., New York, 1961.
[152] Reinmutch, *Anal. Chem.* **36** (1964) 211R.
[153] Berg, *Naturweissenschafter* **47** (1960) 320; **49** (1962) 11.
[154] Yeager, *Transactions of the Symposium on Electrode Processes, Philadelphia, 1959*, Ed. Yeager, p. 145, John Wiley and Sons, Inc., New York, 1961.
[155] Gerischer and Krause, Z. *Physik. Chem.* **10** (1957) 264.
[156] Warner and Schuldiner, *J. Electrochem. Soc.* **111** (1964) 992.
[157] Anson and Schultz, *Anal. Chem.* **35** (1963) 114.
[158] Giner, *Electrochim. Acta* **4** (1961) 42.
[159] Testa and Reinmuth, *Anal. Chem.* **32** (1960) 1512.
[160] Dapo and Mann, *ibid.* **35** (1963) 677.
[161] Hale, *J. Electroanal. Chem.* **8** (1964) 121.
[162] Sevcik, *Coll. Czech. Chem. Comm.* **13** (1948) 349.
[163] Will, Deutscher Patentant, Pat. applied for W23858 IX/421 (1958).
[164] Julliard and Schalit, *J. Electrochem. Soc.* **110** (1963) 1002.
[165] Vielstich, Z. *Instrumentenkunde* **71** (1963) 29.
[166] Hoare, *Electrochim. Acta* **9** (1964) 599.
[167] Srinivasan and Gileadi, *Electrochim. Acta* **11** (1966) 321.
[168] Conway, *J. Electroanal. Chem.* **8** (1964) 486.
[169] Conway, Gileadi, and Angerstein-Kozlowska, *J. Electrochem. Soc.* **112** (1965) 341.
[170] Blomgren, Bockris, and Jesch, *J. Phys. Chem.* **65** (1961) 2000.
[171] Bockris, Müller, and Gileadi, *Electrochim. Acta*, in press.
[172] Bockris, Argade, and Gileadi, *J. Phys. Chem.* **70** (1966) 2044.
[173] Brummer and Ford, Technical Memorandum No. 14, Tyco Laboratories, Inc., August, 1964, Contract No. NONR 3765(00).
[174] Gilman, *J. Phys. Chem.* **66** (1962), 2657; **67** (1963) 78.
[175] Eischens and Pliskin, *Advan. Catalysis* **10** (1958) 18.
[176] Brummer, *J. Phys. Chem.* **69** (1965) 562.
[177] Fleischmann, Johnson, and Kuhn, *J. Electrochem. Soc.* **111** (1964) 602.
[178] Breiter, *Electrochim. Acta* **8** (1963) 447, 457.
[179] Temkin, *Zh. Fiz. Khim. SSSR* **15** (1941) 296.
[180] Bagotzky and Vasilyev, *Electrochim. Acta* **9** (1964) 869.
[181] Hayward and Trapnell, *Chemisorption*, Butterworths, London 1964.
[182] Bond, *Catalysis by Metals*, Academic Press Inc., New York 1962.
[183] Twigg and Rideal, *Proc. Roy. Soc. London, Ser. A* **171** (1939) 55.
[184] Conn and Twigg, *Proc. Roy. Soc. London, Ser. A* **171** (1939) 70.
[185] Beeck, Smith, and Wheeler, *Proc. Roy. Soc. London, Ser. A* **177** (1940) 62.
[186] Twigg and Rideal, *Trans. Faraday Soc.* **36** (1940) 533.
[187] Beeck, *Disc. Faraday Soc.* **8** (1958) 118; *Rev. Mod. Phys.* **17** (1945) 61.
[188] Trapnell, *Trans. Faraday Soc.* **48** (1952) 160.
[189] Jenkins and Rideal, *J. Chem. Soc.* (1955) 2490, 2496.
[190] Selwood, *J. Amer. Chem. Soc.* **79** (1957) 3346.
[191] Arthur and Hanson, Institute for Atomic Research and Dept. of Chem., Iowa State U. Ames, Iowa, Contribution No. 1063.
[192] Coulson, *Valence*, Oxford University Press, London 1961.
[193] Johnson, Wroblowa, and Bockris, *J. Electrochem. Soc.* **111** (1964) 863.

[194] Eischens, *Science* **146** (1964) 486.

[195] Little, Sheppard, and Yates, *Proc. Roy. Soc. London, Ser. A* **259** (1960) 242.

[196] Niedrach, *J. Electrochem. Soc.* **111** (1964) 1309.

[197] Wroblowa and Green, *Electrochim. Acta* **8** (1963) 679.

[198] Hieland, Gileadi, and Bockris, *J. Phys. Chem.* **70** (1966) 1207.

[199] Makrides, Brummer, and Donit, Tyco Laboratories, Inc., Walthan, Mass., Second Interim Technical Report, May, 1964–Oct., 1964. Contract DA44-009 AMC 410 (T).

[200] Flannery, Aronowitz, and Walker, American Oil Company, Research and Development Dept., Whiting, Ind., Progress Report No. 1, March, 1965, Contract DA-49-186-AMC-167(X).

[201] Rao, Damjanovic, and Bockris, *J. Phys. Chem.* **67** (1963) 2508.

[202] Bockris, Piersma, Gileadi, and Cahan, *J. Electroanal. Chem.* **7** (1964) 417.

[203] Gilman, *J. Phys. Chem.* **68** (1964) 70.

[204] Bold and Breiter, *Electrochim. Acta* **5** (1960) 145.

[205] Feldberg, Enke, and Bricker, *J. Electrochem. Soc.* **110** (1963) 826.

[206] Liang and Franklin, *Electrochim. Acta* **9** (1964) 517.

[207] Buck and Griffith, *J. Electrochem. Soc.* **109** (1962) 1005.

[208] Schwabe, *Z. Elektrochem.* **61** (1957) 744.

[209] Munsen, General Electric Research Lab., Schnectady, N.Y., Summary Report No. 1, March 31, 1962, Contract DA-44-009-ENG-4853.

[210] Gottlieb, *J. Electrochem. Soc.* **111** (1964) 465.

[211] Slott, Ph.D. Thesis, M.I.T. (1963).

[212] Oxley, Johnson, and Buzalski, *Electrochim. Acta* **9** (1964) 897.

[213] Frumkin and Podlovchenko, *Dokl. Akad. Nauk. SSSR* **150** (1963), 349.

[214] Podlovchenko, Petrii, and Frumkin, *Dokl. Akad. Nauk. SSSR* **153** (1963) 379.

[215] Giner, *Electrochim. Acta* **9** (1964) 63.

[216] Kutscher and Vielstich, *Electrochim. Acta* **8** (1963) 985.

[217] Flannery *et al.*, American Oil Company Research and Development Dept. Whiting, Ind., Final Report, Sept., 1961–Oct., 1964, Contract DA-11-022-ORD-4023.

[218] Vasilyev and Bogotzky, *Dokl. Akad. Nauk. SSSR* **148** (1963) 132.

[219] Conway and Gileadi, *Trans. Faraday Soc.* **58**, 2493 (1962).

[220] Munsen, *J. Electrochem. Soc.* **111** (1964) 372.

[221] Parsons, *Trans. Faraday Soc.* **54** (1958) 1053.

[222] Mohilner and Delahay, *J. Phys. Chem.* **67** (1963) 588.

[223] Frumkin, *Dokl. Akad. Nauk. SSSR* **154** (1964) 1432.

[224] Bogdanovskii and Shlygin, *Russian J. Phys. Chem.* **33** (1951) 151.

[225] Bogdanovskii and Shlygin, *Zn. Fiz. Khim.* **34** (1960) 57.

[226] Johnson, Wroblowa, and Bockris, *Electrochim. Acta* **9** (1964) 639.

[227] Akuberg, *Z. Anorg. Chem.* **31** (1902) 161.

[228] Graig and Hoffmann, *Nat. Bur. Std. U.S., Circ.* 524 (1953).

[229] Rius, Llopis, and Seinet, *An. Real Soc. Esp. Fis. Quím.* **49** (1953) 447.

[230] Klemenc, *Z. Phys. Chem.* **185A** (1939) 1.

[231] Llopis, Fernandez Biaze, and Guillen, *An. Real Soc. Esp. Fis. Quím.* **52B** (1956) 601.

[232] El Wakkad, Khalafalla, and Sham El Din, *Egypt J. Chem.* **1** (1958) 23.

[233] Anson and Schultz, *Anal. Chem.* **35** (1963) 114.

[234] Johnson, private communication.

[235] Fichter, *J. Electrochem. Soc.* **75** (1939) 309.

[236] Gentile, Smith, Williams, and Driscoll, Monsanto Research Corp., Boston Lab., Report #1, Contract DA 36-039-SC-88945, Jan.–Oct., 1962.

[237] Wilson and Lippincott, *J. Am. Chem. Soc.* **78** (1956) 4290.
[238] Davydov and D'yachko, *Russian J. Phys. Chem.* **37** (1963) 93.
[239] Khomutov and Khachaturyan, *Tr. Mosk. Khim.-Tekhnol. Inst. im. D.I. Mendeleeva* **32** (1961) 207.
[240] Khomutov and Khachaturyan, *Vtorol Sowshchanie po Elektrokhimii Organisheskikh Soedinenii, Tezisy Dokladov,* p. 36, Moscow, 1959.
[241] Glasstone and Hickling, *Trans. Electrochem. Soc.* **75** (1939) 333.
[242] Fioshin, Vasilyev, and Gaginkina, *Dokl. Akad. Nauk SSSR* **135** (1960) 909.
[243] Erschler, *Disc. Faraday Soc.* **1** (1947) 269.
[244] Frumkin, *Trans. Faraday Soc.* **55** (1959) 156.
[245] Kolotyrkin, *Trans. Faraday Soc.* **55** (1959) 455.
[246] MacDonald and Conway, *Proc. Roy. Soc. London, Ser. A* **269** (1962) 419.
[247] Meyer, *J. Electrochem. Soc.* **107** (1960) 847.
[248] Khomutov, *Russian J. Phys. Chem.* **34** (1960) 851.
[249] Khomutov, *Tr. Mosk. Khim-Tekhnol. Inst. im. D.I. Mendeleeva* **32** (1961) 115.
[250] Hayer, *Z. Naturforsch* **4a** (1949) 335.
[251] Bogdanovskii and Shlygin, *Zh. Fiz. Khim.* **33** (1959) 1769.
[252] Leeds, *Ber.* **14** (1881) 977.
[253] Breiter, *Electrochim. Acta* **7** (1962) 533.
[254] Gilman and Breiter, *J. Electrochem. Soc.* **109** (1962) 1099.
[255] Breiter, *J. Electrochem. Soc.* **110** (1963) 449.
[256] Takamura and Minamiyama, *J. Electrochem. Soc.* **112** (1965) 333.
[257] Rightmire, Rourland, Boos, and Beals, *J. Electrochem. Soc.* **111** (1964) 242.
[258] Zeller, *Trans. Electrochem. Soc.* **92** (1947) 335.
[259] Korolev and Shlygin, *Russian J. Phys. Chem.* **36** (1962) 156.
[260] Khomutov and Bystrov, *Russian J. Phys. Chem.* **36** (1962) 1212.
[261] Advances in Chemistry Series 47, Ed. Could, p. 292, American Chemical Society, Washington, D.C., 1965.
[262] Bockris, Stoner, and Gileadi, *J. Electrochem. Soc.* **113** (1966) 585.
[263] Bockris and Huq, *Proc. Roy. Soc.* **237** (1956) 277.
[264] Vetter and Berndt, *Electrochem. Acta* **62** (1958) 378.
[265] Conway, Beatty, and DeMaine, *Electrochim. Acta* **7** (1962) 39.
[266] Argade and Gileadi, in *Electrosorption,* Ed. Gileadi, Plenum Press, New York, 1966, in press.
[267] Kheifets and Krasikov, *Zh. Fiz. Khim.* **31** (1957) 1992.
[268] Chasova, Vasilyev and Bagotzky, *Electrokim. Acta* **1** (1965).
[269] Frumkin, *Electrochim. Acta* **9** (1964) 465.
[270] Hickling and Salt, *Trans. Faraday Soc.* **36** (1940) 1226.
[271] Bockris, *Trans. Faraday Soc.* **43** (1946) 417.
[272] Masing and Lave, *Z. Physik. Chem.* **178** (1936) 1.
[273] Knorr and Schwartz, *Z. Elektrochem.* **40** (1934) 38; *Z. Physik. Chem.* **176** (1936) 161.
[274] St. von Naray-Szabo, *Z. Physik. Chem.* **178** (1937) 355.
[275] Gileadi and Fullenwider, in press.
[276] Busing and Kauzmann, *J. Chem. Phys.* **20** (1952) 1129.
[277] Delahay and Trachtenberg, *J. Am. Chem. Soc.* **79** (1957) 2355.
[278] Ward, *Proc. Roy. Soc. London, Ser. A* **133** (1931) 522.
[279] Hawley and Adams, *J. Electroanal. Chem.* **8** (1964) 163.
[280] Mizoguchi and Adams, *J. Am. Chem. Soc.* **24** (1962) 2058.
[281] Galus and Adams, *ibid.* **24** (1962) 2061.
[282] Galus and Adams, *J. Phys. Chem.* **67** (1963) 862.
[283] Adams, McClure, and Morris, *Anal. Chem.* **30** (1958) 421.

[284] Hedenberg and Freiser, *Anal. Chem.* **25** (1953) 1355.
[285] Penketh, *J. Appl. Chem.* **1** (1957) 512.
[286] Gaylor, Elving, and Conrad, *Anal. Chem.* **25** (1953) 1078.
[287] Julian and Ruby, *J. Am. Chem. Soc.* **72** (1950) 4719.
[288] Packer and Adams, *Anal. Chem.* **28** (1956) 828.
[289] Voorhies and Adams, *Anal. Chem.* **30** (1958) 346.
[290] Lee and Adams, *Anal. Chem.* **34** (1962) 1507.
[291] Loveland and Dimeler, *Anal. Chem.* **33** (1961) 1196.
[292] Pysh and Yang, *J. Am. Chem. Soc.* **85** (1963) 2124.
[293] Suatoni, Snyder, and Clark, *Anal. Chem.* **33** (1961) 1894.
[294] Lu-an, Vasilyev, and Bagotzky, *Zh. Fiz. Khim.* **38** (1964) 205.
[295] Frumkin in 1928 [cf. Vestnik Moscow Univ., Chem. *Ser.* **9** (1952) 37].
[296] Antropov, *Zh. Fiz. Khim.* **25** (1951) 1494, cf. also *Kinetics of Electrode Processes and the Null Points of Metals*, Antropov, published by the Council of Scientific and Industrial Research, New Delhi, 1960.
[297] Frumkin, *Z. Elektrochem.* **59** (1955) 807.
[298] Blomgren and Bockris, *J. Phys. Chem.* **63** (1959) 1475.
[299] Swinkels, Ph.D. Thesis, University of Pennsylvania (1963).
[300] Conway, Chairman's Address, First Australian Conference on Electrochemistry, 1963.
[301] Balandin, *Z. Physik. Chem.* **B2** (1929), 289; **B3** (1929) 167.
[302] Cunningham and Gwathmey, *Advances in Catalysis*, Vol. 9, p. 25, Academic Press, New York, 1957.
[303] Hill, *J. Chem. Phys.* **16** (1948) 181; *Advances in Catalysis*, Vol. 4, p. 236, Academic Press, New York, 1952.
[304] Dowden, *Chemisorption*, Ed. Garner, Butterworths, London, 1957.
[305] Adlhart and Hener, *Fuel Cell Catalysis*, Final Report, Contract No. DA 36-039, SC 90691, U.S. Army Electronics Research and Development Laboratories.
[306] Breiter, *Electrochim. Acta* **8** (1963) 973.
[307] Conway and Bockris, *J. Chem. Phys.* **26** (1957) 532.
[308] Bockris and Wroblowa, *J. Electroanal. Chem.* **7** (1964) 428.
[309] Drazic and Bockris, *Electrochim. Acta* **7** (1962) 293.
[310] Glasstone, Laidler, and Eyring, *The Theory of Rate Processes*, McGraw Hill, New York, 1941.
[311] Wroblowa, Kuhn, and Bockris, in preparation.
[312] Hirota, Kumata, and Asai, *Nippon Kazaku Zasshi* **80** (1959) 701.

3

Ionic and Electronic Currents at High Fields in Anodic Oxide Films

W. S. Goruk, L. Young, and F. G. R. Zobel

I. INTRODUCTION

The ionic and electronic conductivity of anodic oxide films will be discussed in terms of the behavior of films on tantalum, with only brief mention of films on silicon, niobium, aluminum, and a few other materials.

For those unfamiliar with anodic oxide films we note that the kind of anodic films to which we are limiting ourselves are flat, parallel-sided slabs, $10-10^4$ Å in thickness and made of glassy oxide material. They are grown by making the metal the anode in a suitable solution. Unless special precautions are taken, a thin film of the order of a few tens of angstroms is present before current is applied. When current is applied, an electric field of magnitude 10^6-10^7 V/cm is set up in the oxide, and this causes either metal or oxygen ions, or, as appears to be the case with tantalum, both types of ion, to move through the film and fresh oxide to form.

The field in the oxide is not directly obtained from the measured voltages because one has the usual problem of electrochemical systems involving differences of inner potential between phases of different composition. With reasonably large voltages this problem is not acute, but the accuracy of recent work is such that the proper procedure must be adopted in kinetic studies even with voltages of 100–200 V. Reference below to the field strength in the oxide will usually imply an average quantity; the overpotential for the oxide-producing reaction divided by the thickness of the film.

With tantalum at ordinary experimental rates of growth, the electronic current through the film during the growth process is negligible; however, with silicon, which requires fields of about 2×10^7 V/cm for appreciable ionic currents, the electronic current may greatly exceed the ionic current. Experimenters have generally assumed that the ionic and electronic currents are independent, so that the latter may be neglected in studying the former.

Because of the thinness of the films, ionic and electronic currents may be studied at these high field strengths with current densities of up to tens of mA/cm^2 and more without having to use pulses to avoid heating.

The central problem in the study of ionic conduction is to discover the details of the atomic transport processes involved in the growth of films. This will be discussed in the first part of this review. The electronic conductivity is of considerable theoretical interest and is of great practical importance for microelectronic devices. We discuss the system in which a thin metal counter-electrode replaces the electrolyte solution in which the oxide was made. Thermionic and field assisted emission, tunneling processes, impurity band conduction, and space-charge limited currents, have to be considered. We shall draw on results for oxide films made by other processes, such as evaporation and thermally promoted reaction with oxygen.

The theories of the electronic and ionic currents have some features in common. One may formulate models in which the current is limited by the injection into the film from the contacts of positively or negatively charged carriers, or one may consider an equilibrium state to exist across either or both interfaces. One may postulate space-charge limited currents, trapping, and recombination processes. One of the chief differences between the ionic and the electronic currents is that the average velocity of the ions is approximately exponentially dependent on the field for fields which produce experimentally observable ionic currents, whereas the average velocity of electrons is linearly dependent on the field at low fields with different types of nonlinearity at high fields.

Prior to 1953, it was widely thought that the process of ionic conduction was well understood in principle. The main uncertainty appeared to be whether the injection of ions into the film or their transport through the film controlled the relation between field and current. Since 1953, investigations have shown that transport through the film probably controls the kinetics, but that the basic transport process is more complicated than had been thought. Several models have been proposed, but they have failed to account for new experiments. The assembly of known effects is now considerably enlarged, and there is some disagreement on details of the kinetics. No generally accepted theory exists in terms of the details of the atomic processes.

We have, for reasons of clarity, rejected an historical approach, and we have not aimed at a comprehensive account of the literature. A detailed account of the properties of anodic oxide films on a variety of materials was published recently by Young,[1] and Vermilyea[2] has also reviewed the subject.

II. STRUCTURE OF THE FILMS

The structure of the films and the nature of the film materials are basic to the theory of the ionic and electronic transport processes.

The films on tantalum are usually described as glassy or amorphous. Broad halos are produced by electron and X-ray diffraction. The structures of the crystalline forms of the bulk Ta_2O_5 still, so far as we can determine, have not been worked out. Calvert and Draper[3] have proposed that β-Ta_2O_5, which is the form stable below about 1300°C, and which is also, they believe, the more closely related to the material of the anodic films, consists of chains of tantalum and oxygen ions, parallel to the c axis, with various degrees of disorder possible in the plane at right angles.* Following the emphasis which was placed by Vermilyea[4] on the glassy nature of the films, it is more usual to regard the structure as an aperiodic three-dimensional network of tantalum and oxygen ions, with the mobile tantalum ions occupying what are known in glass technology as network-modifying positions. Here, one must attempt to distinguish between tantalum ions firmly held in place in the lattice, and ions loosely resting in interstices, and, therefore, more available to produce current. It is debatable whether there are two distinct distributions of sites or whether interstitial and network sites are just two ends of a single distribution. The ionic and the electronic conductivity change continuously and by large

*See also Lehovec and Dreiner, J. Less-Common Metals 7(6) (1964) 397.

amounts under the conditions the film is made, or to which it is later subjected. These changes usually have been thought to be due to a change in the concentration of interstitial ions (that is, to a shift of ions between network and interstitial positions) and to a varying concentration of oxygen ion vacancies. But, as suggested above, the metal ions should possibly be considered as redistributing themselves over a single continuous distribution of sites.

The parallel-sided flat-slab nature of the films is different from what is often found with films formed by thermal oxidation at higher temperatures. Growth features, such as surface contours which reveal the outline of the individual crystals of oxide below, are absent in the amorphous anodic films and present in the crystalline films grown by exposure to oxygen at moderately high temperatures. Whisker-like protuberances are also absent in the anodic films—at least, have not been reported. However, structural features are present in the form of gross defects of one sort or another. For example, patches of well-crystallized oxide are produced under some conditions (field-recrystallization), and surprisingly, grow thicker than the surrounding amorphous oxide.[5,6] A category of defect which is trivial enough in itself, but which has nuisance value in the study of electronic conduction and dielectric breakdown, includes those defects which involve a physical break in the film, or other features such as local nonstoichiometry, or impurity doping, which allow electronic current to flow easily. It has been known for many years[7] that if the oxide-coated metal is made cathodic in a solution containing electrolytically depositable metal ions, metal is deposited only here and there on the oxide surface. This indicates weak places where electrons may be easily exchanged with ions in solution. Such places may be centimeters apart with well-prepared surfaces. The presence of actual breaks in the films is shown by capacitance measurements in solutions of different conductivity.[8]

Vermilyea[9] has recently observed a peculiar and characteristic form of defect (associated with impurities in the metal) which may be present in great numbers in films of certain thicknesses on metal of ordinary purity. These defects do not necessarily lead to easy conduction. They consist of a region where the growth of the oxide appears to have become inhibited. This results in a protuberance in the metal and a depression in the outer surface of the oxide

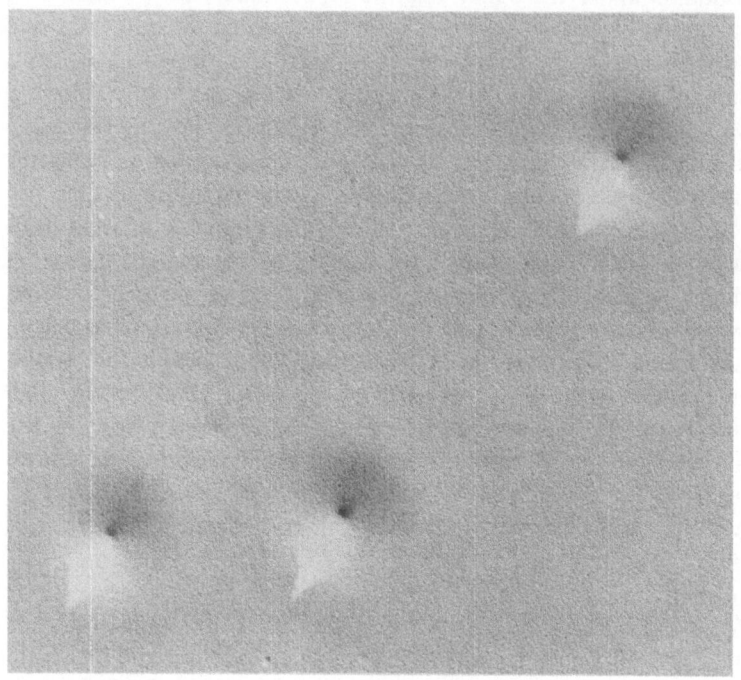

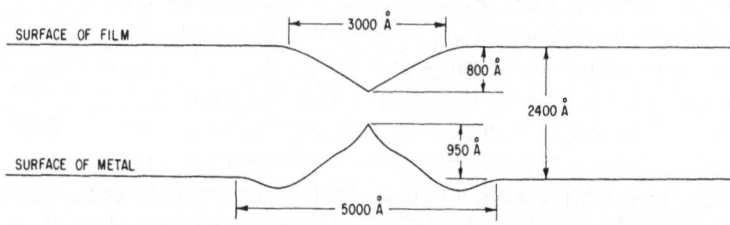

Figure 1. Surface of chemically polished tantalum after stripping off a 3100 Å oxide film. Dimensions of flaws of this type are shown in the diagram. Height of the elevations was estimated from shadow lengths [from Vermilyea, *J. Electrochem. Soc.* **110** (1963) 250].

(Figure 1). The film may even become detached as the surrounding oxide grows thicker. The defects occur at concentrations of $10^4 \, cm^{-2}$ with the purest metal and up to $10^{10} \, cm^{-2}$ with very impure metal. It would appear that they are fewer with chemically polished metal than with "as-received" metal. To avoid confusion we note that usually the word "defects" will refer to interstitial ions and the like.

The stresses in the films have recently been investigated by Vermilyea,[10] who observed the deflection of strips of metal when oxide was formed on only one side. An old misconception is, that if, as here, the oxide occupies more space than the metal from which it is formed, the oxide should be in compression. With a flat surface this is not so, since the atoms move one by one, and any stresses will be due to a subtle process of local adjustments to the movement of ions and to changes in shape where surface contours are smoothed by the growth of the films.[11,12] Vermilyea found that the films are mostly under tensile stress.

The changes in the film material with the current density, field, and temperature at which the films are made (in dilute, aqueous solution) and the changes which occur on subsequent annealing at temperatures of up to 200°C or so, represent large proportional changes in the concentration of mobile ions, that is, according to the usual theory, in the concentration of defects (interstitial metal ions and anion and cation vacancies). However, the absolute variations in terms of numbers of defects are probably small. Thus the variations in properties, which are not directly functions of defect concentrations, such as refractive index[15] and density,[16] are of the order of 1 %—not large for ordinary formation conditions—whereas the variations in those properties, such as ionic conductivity,[4,13] electronic conductivity, and rate of dissolution in HF,[14] which are directly dependent on the concentration of lattice defects, are easily detected. More information is undoubtedly needed on these variations.

Atomic rearrangements occur quite readily in these films at room temperatures. Thus, when the formation current is cut off and the field reduced to zero, the small signal differential capacitance and dielectric losses decrease by appreciable amounts within a few minutes,[17] the change in the dielectric constant being of the order of 1 or 2 %. The decay in the ionic conductivity, as measured by the value of the ionic current which is observed immediately following the reapplication of a field, has been studied recently[18] following earlier work by Dewald[19] and Vermilyea.[4] It falls rapidly at first—by appreciable amounts in a few tens of milliseconds—and is still decreasing after some hours (Figure 2). These

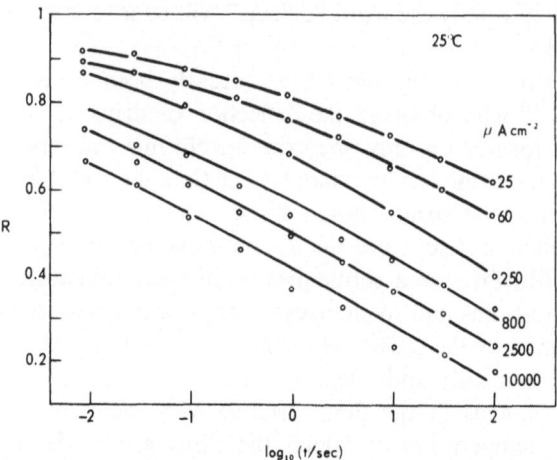

Figure 2. Fraction R of original ionic current observed on reapplying the same field after various time intervals at zero field for films held previously at constant fields giving the currents indicated. R represents the proportion of mobile ions which have not moved into sites where they do not contribute to the ionic current [from Young, J. Electrochem. Soc. **111** (1964) 1289].

effects are evidence for a redistribution of ions among available sites and for a wide variation in the properties of available sites. One supposes that ions left in interstitial-like positions when the current is cut off shift around to adopt the lowest energy sites available in their vicinity. When the field is reapplied, the ions which have moved into lower energy sites are no longer so free to move. This local movement of ions, occurring after the cessation of the ordered macroscopic flow which constitutes the ionic current density, is an example of a local rearrangement process such as occurs in a variety of recrystallization and other solid-state reactions. The local process may here be compared with the more easily studied macroscopic process. The disappearance of the mobile ions may be treated as due to their transport under the action of local electric fields associated with the charges on interstitial ions and vacant cation sites, although this is, as follows from the above discussion, an oversimplification. The variation of relaxation times will then depend on the distance to the nearest vacant cation site. The experiments (Figure 2) confirm the general

picture. For example, the higher the field during the initial formation at a given temperature, the faster the proportional loss of
mobile ions ("recombination of Frenkel defects"). This is to be
expected (if there are more defects, the higher the field) since the
components of the defects are then closer together, and do not have
so far to go to recombine.

Optical measurements[15,21] and also measurements of capacitance, field, charge-to-form,[22] rates of dissolution in HF, and mass,
indicate a remarkable lack of dependence of film properties on
total thickness and a remarkable degree of homogeneity through
the thickness of the film. One exception[15] is that certain optical
effects indicated a thin outer layer of light-absorbing oxide,
possibly a transition layer a few Å thick at the oxide solution interface. Heavens and Kelly[23] reported measurements which indicated
a very inhomogeneous film and large variation with thickness, but
these measurements were made on thin evaporated films which
were probably impure. Using ellipsometry, Claussen[24] has found
that variation with film thickness occurs with films formed on
sputtered tantalum films, the precise behavior depending on the
technique of sputtering. (See also the ellipsometry data in section
III.2.xii.) (See *Notes added in proof.*)

III. IONIC CURRENT

1. Model of Ionic Transport Process at High Fields

Before becoming involved in the complexities of the behavior of
the anodic films, it is convenient to discuss the model of the basic
transport processes which is involved in ionic conduction at high
field strengths. The classical model for ionic conduction in crystalline solids, as presented by Frenkel[25] in relation to the deviations
from Ohm's Law observed by Poole and others, is as follows: First,
the mobile species are defects. Considering the movement of interstitial ions, we note that the mobile ion acquires enough energy by
thermal fluctuation to jump to a neighboring site, whereupon it is
immediately deactivated, since most collisions suffered by an ion
which has a large energy will result in loss of energy. The ion
remains oscillating in this new site, with an average energy of the
order kT and at a frequency around 10^{13} cps, until it next acquires
the energy, usually much greater than kT, required for a further

jump over a potential-energy barrier. The field gives an assisting push, so that a given ion moves with the field more often than against it. If the work done by the field E, as the ion with charge q moves the distance a from the potential energy minimum to the potential energy maximum, is much greater than kT, movement against the field will be negligible, and the mean velocity will be $2av \exp[-W(E)/kT]$, where v is a vibration frequency, $\exp[-W(E)/kT]$ is the probability that a linear harmonic oscillator has energy $W(E)$, and the dependence of $W(E)$ on E is written $W_0 - qaE$. This basic model may be elaborated by attempting to allow for notions of collective motion of nearby ions.[26] An ion is most likely to jump when its neighbors happen to move in an ordered way such as to open up a passage for it. The model may be described in terms of the transition-state theory. One may also try to improve on the purely statistical treatment of the problem by considering the mechanism by which the ion acquires its large energy and whether, in view of the rapid rate at which an ion acquires energy from the field while in motion, the ions do not move through several sites before sticking. Some of these points will be discussed below. It is convenient to discuss another effect immediately. It has recently been asserted,[16,17] that one would in general expect the activation energy to have an expansion in powers of E, i.e., $W(E) = W_0 + \alpha E + \beta E^2 + \ldots$ so that given sufficiently accurate measurements, one would require the retention of the quadratic term as well as the ordinary linear term. It was suggested that the present experimental results do in fact require this term, with α negative and β positive. Various physical processes could be responsible for a quadratic term. The compression of the oxide owing to condenser pressure or to electrostriction in an amorphous material will be proportional to E^2, and appears likely to give an appreciable effect with positive β.[28] Other contributions may also be significant. If the ion is held in the "site" by Coulombic attraction and the "sites" are widely spaced in channels, the Schottky-type law would be obeyed. The $E^{\frac{1}{2}}$ dependence could, as will be shown in discussion of the experimental results, be described with the quadratic term over the small range of field which is accessible. An effect of a rather different sort would occur where a statistical distribution in site parameters occurs and a series of barriers must be taken in turn by each ion in moving through the crystal. This case was discussed in detail elsewhere.[16,1] The condenser pressure at these fields, as was noted by Güntherschulze and Betz many years ago,[7] is hundreds of atmospheres. It is well known[29] that a reduction of Ohmic, low-field ionic conductivity is expected and observed because of the compression of the lattice with directly applied pressures. Calculations (unpublished) based on activation volumes for ohmic ionic conduction in doped halide crystals[30] suggest too small an effect to explain the observations discussed below with tantalum pentoxide, but the latter has a more open structure than the halides and is hence probably more compressible. Experiments on anodic films with directly applied pressures have been irreproducible because of current fluctuations of a somewhat mysterious origin, possibly associated with defects in the films. It is emphasized that although the form of equation was first introduced to describe experimental results, the suggestion is that this form is expected theoretically. In electrochemistry, curved Tafel lines have usually been assumed to indicate a mixture of mechanisms. This assumption is not always correct.

When the field contribution is written qaE, there is some question as to whether the effective field should not be some form of cavity field. Also, in the kinetic theory one might imagine that the activation distance would be larger than the distance from minimum to maximum potential energy, since the ion presumably starts its run from the other side of the potential-energy minimum. Finally, the effective charge on an ion may not correspond to that associated with the valency in the compound concerned.[31]

The usual concept of a time-independent potential energy-distance relation discounts some important features of the transport process. The energy of an ion will depend on phase relations with the motion of neighboring ions. Such potential-energy distance relations are partly analogous to the periodic field in the one-electron theory of electronic states in crystals. What one has in the present case is a particular set of configurations of position and momenta of the mobile ion and its neighbors which favor motion. The assembly in this favorable state constitutes the transition state whose concentration is given in equilibrium by an expression which includes the factor $\exp(-\Delta G^+/kT)$. The rate of jumping is then obtained by including a frequency factor for the rate of decomposition of the transition state, and the concentration of mobile ions may be estimated.[18,28] An entropy factor, $\exp(\Delta S^*/k)$ where ΔS^* is positive, must be included to reduce the estimate to a reasonable level. The same effect in low-field ionic conductivity is ascribed to the dependence of activation energy on temperature due to thermal expansion of the lattice.[29]

In attempting to formulate the process by which ions acquire the energy necessary to move, the theory of focused collisions is of interest.[32] This is a theory of the sputtering process in which positive ion bombardment of, say, a tantalum cathode results in tantalum atoms or ions flying off into space. The energy supplied by the positive ions is passed from atom to atom by successive collisions until the energy packet reaches a free surface. The last atom in line has no atom to receive its kinetic energy. The energy packet may be focused into a line of atoms and may travel without much attenuation in crystalline material. With thin foils, atoms may be sputtered from the surface opposite to that bombarded. This unlikely process works very efficiently, and provides justification for considering the possibility of similar effects in the present systems.

Finally, a few numbers may clarify the picture. A tantalum ion with a kinetic energy of 1 eV, which is the order of magnitude of the activation energies with which we are concerned, has a velocity of 10^5 cm/sec. Starting with this minimum velocity for jumping, the ion is supposed to be just braked to a standstill at the top of the barrier, that is, in a distance of about 1.5 Å. The experimental range of fields is small, because of the exponential dependence of current on field. With tantalum, it is about 4 to 7×10^6 V/cm. For 5×10^6 V/cm the Ta^{5+} ions have 0.25 eV work done on them in traveling 1 Å, which requires 10^{-13} sec at this speed. This time is just of the order of the period of atomic vibrations (optical modes).

2. Experimental Data on Ionic Conduction in Films on Tantalum

(i) Thickness and Field

The field strength for a given ionic current appears to have been established (using unequivocal optical techniques to obtain the thickness) to be constant to within about 1% from 200 to several thousand angstroms.[16,20,33,34] There is a considerable body of information using less reliable techniques to determine thickness (in particular, capacitance) which at least has not indicated any definite change to much smaller thicknesses (for references see Ref. 1). Further work using ellipsometry is required to settle the matter with very thin films. At constant ionic current this implies that the thickness and potential increase linearly with time.

This constancy of the field implies that either electroneutrality prevails in the bulk of the film or the concentration of ions is not controlled by electroneutrality, but that the films are so thin, and the density of carriers so low, that the space charge due to the carriers causes negligible change in field across the film. The first case was that considered implicitly by Verwey,[35] who was the first to apply Frenkel's theory to these systems. The second was proposed by Mott,[36] and was discussed by Mott and Gurney,[37] and by Cabrera and Mott.[38]

(ii) Dependence of Velocity of Ions on Field

A consideration of the steady-state (i.e., constant-field) relation between current, field, and temperature is postponed, since the behavior following changes in the current and the field seems to imply that the concentration of mobile ions (or their distribution among the available sites), as well as their velocity, varies with the field under steady-state conditions. Therefore, the steady-state kinetics are not simple to interpret. Instead, we consider what happens when the field is changed rapidly from a given initial value, under which steady-state conditions have been set up, and the immediate value of the ionic current at the new field is observed. It is assumed that in this case the change in current is due solely to a change in the mean velocity of whatever mobile species are present, no change occurring in their concentration. Vermilyea[4] first showed that under these conditions the relation $(\partial E/\partial \log I)_{transient} \propto T$ which is expected from the classic theory is not observed, this quantity actually decreasing with increasing temperature under the conditions used (Figure 3). Later measurements by Young[27] (Figure 4a) could be described by the substitution $W(E) = W_0 - \alpha E + \beta E^2$ in the equation for ion mobility. This implies that $(\partial E/\partial \log I)_{transient}$ depends on E at constant T. The evidence, discussed later, that oxygen ion mobility contributes an appreciable portion of the observed ionic current, leads one to consider whether this could account for the anomaly. However, if we write

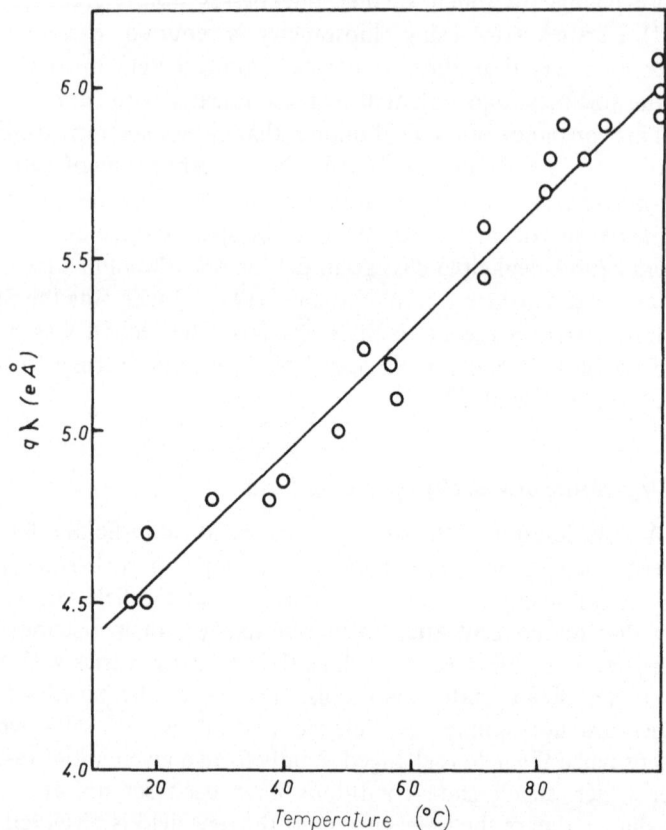

Figure 3. Variation of $q\lambda$ [ie., of $kT/(\partial E/\partial \log i)$] with temperature for suddenly changed field. According to classical theories, $q\lambda$ would be constant. Thickness is based on a density of 8.74 g/cm^3 [from Vermilyea, *J. Electrochem. Soc.* **104** (1957) 427].

$$I^+ = q^+ 2a^+ n^+ v^+ \exp\left[-\frac{W^+(E)}{kT}\right]$$

$$I^- = q^- 2a^- n^- v^- \exp\left[-\frac{W^-(E)}{kT}\right]$$

with $I = I^+ + I^-$ and an electroneutrality condition in the form $n^+ q^+ = n^- q^-$, and assuming the ordinary $W(E) = W_0 - qaE$, we find two linear portions in a $\log I$ *vs.* E plot with the curvature in

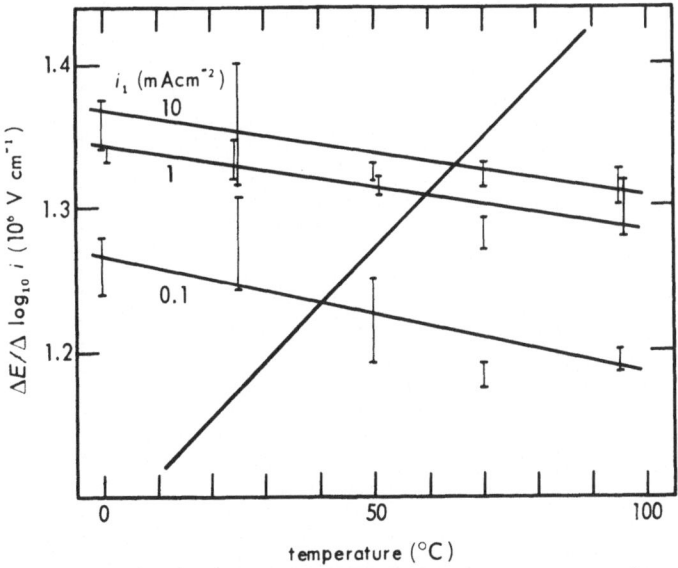

Figure 4a. $\Delta E/\Delta \log_{10} i$ vs. temperature for sudden changes of E with the fields initially brought into steady state at constant fields giving currents shown. The diagonal line with positive slope shows the temperature dependence expected on the classical theory [from Young, *Proc. Roy. Soc.* **263A** (1961) 395].

the transition region, where both I^+ and I^- are comparable, in the opposite sense to that observed. The same result would be found for injection control.

A model was mentioned above in which the ions move freely along channels in the oxide except where they become trapped by Coulombic attraction for local excess negative charges. This model gives a Schottky-type law. The experimental data were fitted by the formula

$$\exp\left[\frac{(-\alpha E + \beta E^2)}{kT}\right]$$

The curve given by this formula plotted on the Schottky plot ($\log I/I_0$ vs. $E^{\frac{1}{2}}$) is almost a straight line (Figure 4b). The slope is predicted theoretically to be

$$\frac{q_2 q_1^{\frac{3}{2}}}{(\varepsilon\varepsilon_0\pi)^{\frac{1}{2}}} \quad \text{(mks)}$$

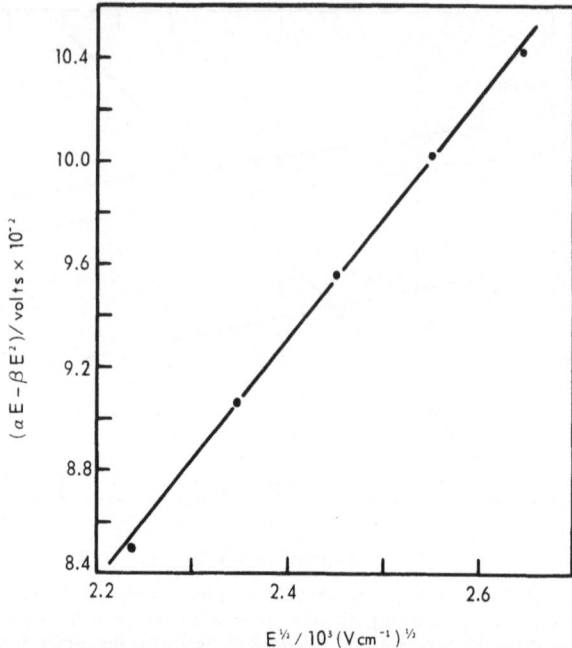

Figure 4b. Points calculated from an expression fitted to experimental data for suddenly changed fields. The plot shows that the velocity of the ions could be written as proportional to $\exp(\beta E^{\frac{1}{4}})$ [data from Young, *Proc. Roy. Soc.* **263A** (1961) 395].

where $q_2 = z_2 e$ is the magnitude of the charge on the mobile tantalum ion, and $q_1 = z_1 e$ is the charge magnitude on the trapping centers. From the experimental slope, assuming $\varepsilon = 28$, with $z_2 = 5$, we obtain $z_1 = 0.1$. For $z_1 = z_2$, both are 1.4. However, one should not jump to the conclusion that the Coulombic trapping-center model is correct. The situation is that one has several effects which will give curvature of Tafel plots and it is not clear how large are the contributions from each effect. The model must also be consistent with all the qualitative features of the experimental behavior.

(iii) *Time Variation of the Current Following a Sudden Increase in the Field*

If a tantalum electrode is anodized at constant voltage, say at 25°C, until the current has fallen to a low value, and the Ta film is then annealed by immersion in boiling water for a few minutes, and a voltage higher than the original voltage is then reapplied at the original temperature, one observes[27] a current–time variation like that shown in Figure 5. Omission of the annealing results in similar but less exaggerated effects. The time scale is compressed when the field is increased. There is an initial bump which could readily be interpreted as a condenser-charging process, except that the time constant is much greater than that calculated from the small signal differential capacitance at low-bias fields. The current then builds up in accelerating fashion. In the specific experiments

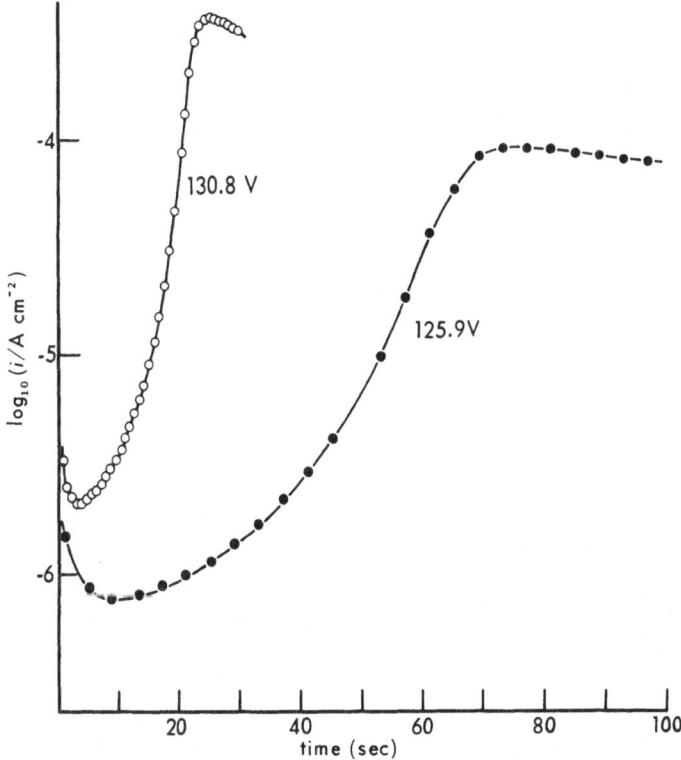

Figure 5. Current as a function of time after application at 0°C of the voltages shown to films held at 100 volts for 3.5 hr at 22°C and annealed at 100°C for 5 min [from Young, *Proc. Roy. Soc.* **263A** (1961) 395].

reported, which covered, admittedly, only a narrow range of conditions, a rather simple law was found: $di/dt \propto i^2$. The proportionality constant was strongly dependent on the field. After this autocatalytic build-up, the current first levels off, and then falls. This fall takes place because as the oxide thickness increases due to the passage of current, the field at constant voltage falls.

This accelerating build up of the current is decisively against the original form of the theory of Bean, Fisher, and Vermilyea[40] which had appeared to offer the basis of an explanation of the kinetics. In this theory, less some complications and as clarified by Dewald,[41] it is supposed that the concentration of interstitial ions is controlled through the condition of electroneutrality by the concentration of vacant cation sites left when the metal ions are removed. These vacant sites are supposed to have a negative charge associated with them because of a local excess of oxygen ions. Frenkel defects were supposed to be produced by the high field by a process analogous to ordinary high-field conduction, giving a rate proportional to $\exp[-(W - qaE)/kT]$. The defects were supposed to be destroyed by the capture by vacancies of interstitial ions moving through the film, giving a rate $I\sigma n$, where I is the current, σ is the capture cross section, and n is the concentration of vacancies. Thus, according to this theory, one would interpret the effect as due to a build-up of Frenkel defects, but the rate of build-up of current should be greatest at the beginning and should fall off as the concentration of defects increases.

The obvious way of trying to modify the theory to fit the above observation is by supposing that the momentum supplied to the lattice by the moving ions is somehow effective in creating defects. Two suggestions may be made to explain the dependence of the rate of increase of the current on the square of the current: First, one might assume that the lattice ions which are particularly liable to be ejected are those at the ends of chains, or near where other ions have already been ejected, and that the concentration of such ions is proportional to the number of ions already moved into interstitial sites. If we further assume that the chance of a given ion being liberated is proportional to the number of jumps being made (that is to the current), each jump being supposed to give one pulse of energy to the lattice which may be transmitted and help eject an ion, we arrive at the required law. The alternative, proposed in the original paper,[27] was that ions were liberated when two energy packets traveling along lines of atoms meet at an intersection, and that the concentration of energy packets is again proportional to the current.

We may attempt to see how what was put forward above as a possibly more realistic model, in which instead of assuming two types of site we postulate a single distribution of sites among which the ions rearrange themselves when different fields are applied, could account for those observations. For example, one might consider a distribution of sites such that $X(W)\,dW$ is the concentration of sites with energies W to $W + dW$ below a common col height. Let $f(W)$ be a function defining the probability that a site of depth W is occupied by a metal ion. If we suppose that (a) the rate at which ions leave sites is given by the classical picture, and (b) that ions, once in motion, travel a distance L (independent of field), and (c) that ions are trapped randomly (independent of W), we have

$$\frac{df\,X}{dt} = -fXK \exp\left[\frac{qaE - W}{kT}\right] + \frac{I(1 - f)X}{qL \int (1 - f)X\,dW}$$

$$I = qL \int fXK \exp\left[\frac{qaE - W}{kT}\right] dW$$

where K is a constant. It is easily shown for this case, if we also assume that there are many more sites than metal ions, that the distribution is given by

$$f = A \exp\left(\frac{W}{kT}\right)$$

A film annealed at higher temperature would give an initial bump of current.

If one now tries to introduce the idea that ions are ejected from sites because of collisions with other ions by replacing the classical term by one of the form MI^2X, where M may be supposed to depend exponentially on E, we find that $f(W)$ tends to zero in the steady state; all the ions are in motion. One might try to rectify this defect in the model by assuming that a field of given magnitude supplies energy packets of limited size, which can release ions only up to a certain W, so that the concentration of mobile ions depends on E. A detailed investigation of the possibilities would be more worth while if one had more detailed experimental data on the transient behavior.

(iv) Dependence of Current in Steady State on Field and Temperature

The steady-state current was predicted by the theories of Verwey[35] and Mott,[36] based on the classical treatment of ion movement, to be given by an equation of the form

$$I = I_0 \exp\left(-\frac{W}{kT}\right)\exp\left(\frac{qaE}{kT}\right)$$

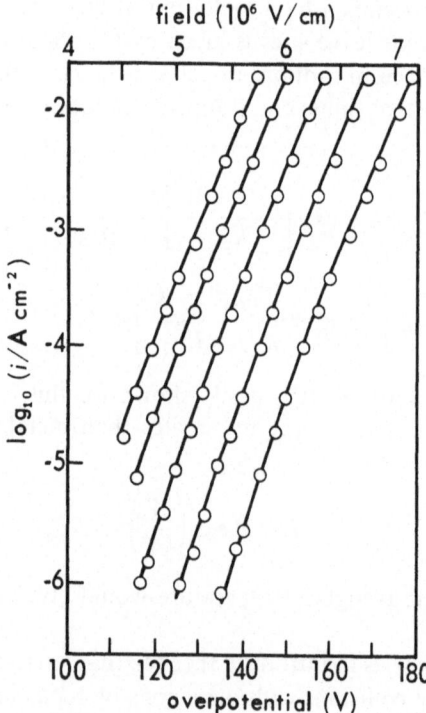

Figure 6a. Overpotential (and hence field = $(V - V_0)/D$) for thickness 2490 Å vs. $\log_{10}$ (ionic current density) in dilute sulfuric acid. Right to left: 0, 25, 50, 75 and 94°C. The curves are the fitted equation [from Young, Proc. Roy. Soc. 258A (1960) 496].

The process was shown to be more complicated than had previously been thought by Vermilyea's discovery[33] of anomalies in the temperature dependence of $\partial E/\partial \log I$. We shall not again review the various unsuccessful explanations which have been put forward to account for the experimental results (see Ref. 1).

It was first thought that $\partial E/\partial \log I$ was independent of E at a given T (as expected from the above theories), but that it was independent of T (instead of being proportional to T as expected). Later experiments[16] were fitted by replacing $W - qaE$ by $W - |\alpha|E + |\beta|E^2$ (Figure 6a). This implies that at a given E the slope $\partial E/\partial \log I$ is proportional to T, but at given I the constants are such that the value of $\partial E/\partial \log I$ is nearly independent of T. The earlier experimental results for $\partial E/\partial \log I$ were essentially for an average slope over a fixed range of current densities.[31,33,39] It seems likely, though the point has not been investigated, that the similar anomaly with niobium[42] could be resolved in the same way.* The above form of the equation fits Vermilyea's later data,[43] though Vermilyea (as already indicated above) put a different interpretation on his results. When allowance is made for the basis of thickness determination (Vermilyea's were based on a value of the bulk density, Young's on an observed value of the refractive index in sodium light) the two sets of data are in excellent numerical agreement (Figure 6b). Recently, some experimental results have been put forward as indicating the temperature dependence expected on the classical theory.[44] The methods which were employed to determine the thickness (charge-to-form and capacity) are less reliable than the optical methods. Also, there is good evidence[16] that the current efficiency is not, as was reported, 11% different at 13.5°C from what it is at 80°C; any difference is of the order of 1%. Experiments in progress by Zobel and Young, using ellipsometry, have confirmed the curvature of the Tafel lines (Figure 6b). Figure 6b is included to show the agreement between recent results. The data are plotted in the Schottky manner ($\log I$ vs. $E^{\frac{1}{2}}$). This form gives quite good linearity, much better than the Tafel plot ($\log I$ vs. E). It is not really meaningful to compare the slope of the Schottky-type plot of steady-state data with the theory for Coulombic traps without introducing some means of explaining the transients. (This argument applies also to Dignam's compar-

*This has now been shown [Young and Zobel, *J. Electrochem. Soc.* **113** (1966) 277].

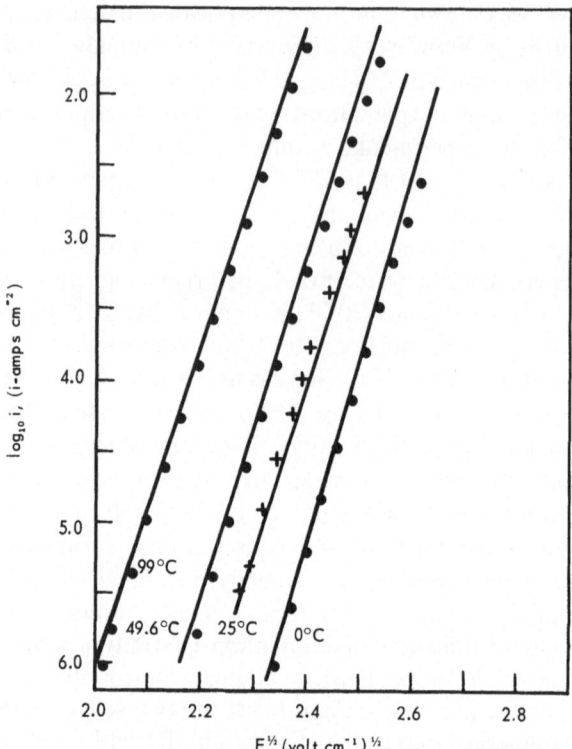

Figure 6b. Schottky plot of data on the steady-state ionic current obtained by different experimenters and brought to the same basis of thickness determination. The smooth lines are calculated from the expression fitted to Young's data (Figure 6a). The dots are experimental data due to Vermilyea,[43] (numerical data supplied by Vermilyea) converted by factor allowing for estimated density of oxide (7.93 × 0.99/8.74). The crosses are preliminary unpublished data obtained by Zobel and converted to the refractive index quoted in Young.[16] [For final version, see Young and Zobel, *J. Electrochem. Soc.* **113** (1966) 277.]

ison[66] of mobility-process theory with the steady-state data.)

In their original paper, Bean, Fisher, and Vermilyea explained Vermilyea's data by assuming the ordinary Frenkel form of equation for the mobility of the interstitial ions. However, for the high-field production of Frenkel defects they assumed a Coulombic attraction between the interstitial ion and the site which it had just left. This Coulombic attraction was superimposed on the periodic potential-energy distance relation. This gave a sudden jump in the activation distance with increasing field as the maximum in the potential energy shifted from between the second and third sites to between the first and second sites. We have suggested elsewhere that any such discontinuity would be smoothed out by the variations from site to site. As Figure 6b shows, the data fit a smooth curve. On the Bean *et al.* theory the slope of this curve is a function of both the process of production of Frenkel defects and the mobility process of these defects.

(v) Evidence That Both Ions Are Mobile

The type of argument which is used to determine which ion is mobile is as follows: If metal ions are mobile, new oxide layers will be built up on the outside of existing layers. If oxygen ions are mobile, new layers will be formed at the oxide/metal interface. If both ions are mobile, oxide may be produced simultaneously at both interfaces or, ion by ion, within the body of the existing oxide.

Most of the earlier work (see Ref. 1) on tantalum and aluminum involved the formation of a marker layer in an electrolyte which produced oxide of different properties. Measurement of the rate of solution of the oxide in HF and of the radioactivity of successively dissolved layers showed where the new layers had been formed. The results, on the whole, indicated metal ion movement, but they were not conclusive. It is well known that in high-temperature oxidation, oxygen migration is sometimes the predominant mechanism. Ligenza and Spitzer[58] and Pliskin and Gnall[59] have shown this by isotopic exchange studies and selective etching, respectively, for silicon oxidized by high-pressure steam and oxygen. Davies *et al.*[49] have studied the high-temperature oxidation of zirconium by steam, and they also found oxygen to be the mobile species. Oxygen is mobile in Ta_2O_5 films heated to 200–300°C. Benjamini, Duffek, Mylroie, and Schulenburg[45] anodized silicon wafers in ethylene glycol, growing two layers, one of which was doped with an ionic phosphorus species. The top layer was dissolved off and the wafers were exposed to high temperature. The remaining oxide was then removed, and the conductivity type of each wafer was determined. A change from *p*- to *n*-type occurred only with the wafers which had been anodized first in the doping solution. This indicated that silicon is the mobile species. However, this experiment again is not conclusive, for if new oxide layers are formed in the middle of the existing oxide, a doping (or nondoping) layer will still remain next to the silicon.

Radioactive-tracer methods have recently been used to study oxygen diffusion in metals.[46] Amsel and Samuel[47] determined the order of atoms in the oxide by the following method: Aluminum was anodized in 3 % ammonium citrate solution, first containing O^{18} in the solvent and the solute, and then O^{16}. The film was stripped from the metal, and bombarded with protons from either side, the resonance from the reaction O^{18} (p, α) N^{15} being measured. It was shown that the distribution of the oxygen isotopes had been conserved to within 60 Å, showing that the metal was mobile. A similar experiment showed that the pre-existing film, 90 Å thick, remained at the metal/oxide interface, its thickness being measured by the resonance technique as 100 ± 30 Å.

In order to trace the metal, a thin layer of aluminum was evaporated onto the tantalum, and the sample was oxidized in a natural oxygen solution to a voltage such that all the aluminum and some of the tantalum would be anodized. The aluminum was traced by means of resonance from the reaction Al^{27} (p, α) Si^{28}. Some mixing of tantalum and aluminum occurred, but the aluminum remained at the tantalum surface. In fact the oxidation of the tantalum was normal, as shown by the interference colors, but the potential was shifted by the amount predicted from the estimated thickness of Al_2O_3. There was no pileup of tantalum ions at the surface. Amsel and Samuel considered the experiment to be inconclusive because Al_2O_3 and Ta_2O_5 have different crystal structures. They postulated that the metal ions move by lattice vacancy diffusion through a stationary oxygen sublattice. Metal ion movement is expected from the smaller ionic radii of the metal ions (O^{2-}, 1.4 Å; Al^{3+}, 0.5 Å; Ta^{5+}, 0.7 Å).

A β-ray spectrometry method has been used by Davies, Pringle, Graham, and Brown[48,49] to determine the depth of inert gas isotope atoms (Xe^{125}) embedded in solids. The Xe^{125} emits monoenergetic conversion electrons, the most intense spectral lines having electron energies of 21.79 keV and 155.2 keV, arising from the K conversions of the 54.96 keV and 188.4 keV nuclear transitions, respectively. The shape of the conversion lines, which was analyzed with a β-ray spectrometer depends on the thickness of material that the electrons must traverse after emission, and this dependence was found by subliming known amounts of aluminum or gold over a Xe^{125} source. It was found that the peak height decreased approximately exponentially with Xe^{125} depth. Under the best conditions the resolution was about 15 Å, or six atom layers. The method was applied to the anodic oxidation of aluminum and tantalum,[50] by initially bombarding the metal foils with Xe^{125}, and then measuring the location of these atoms at various stages during anodization. The effect on the shape of the K 54.96 line and the Xe^{125} distribution is shown in Figure 7 for aluminum and tantalum, respectively. With tantalum anodized at 2 mA/cm^2, the Xe^{125} atoms were found at about the middle of the oxide film indicating that both oxygen and metal migration contribute to the film growth. With aluminum anodized at 0.1 mA/cm^2, the Xe^{125} remained in the outer layers of oxide. This was interpreted as being due to oxygen migration alone. It was found earlier[51] that, under conditions of constant voltage, about 20% of the aluminum consumed in the anodizing process is dissolved by the electrolyte. This seems out of line with other results.

There is some evidence[52] that pores are formed in the outermost oxide layers of films of Al_2O_3 when anodized at low current density in ammonium citrate, and that subsequent film growth occurs underneath this region. Indeed, if this were the case, the Xe^{125} showed that the behavior in aluminum depended markedly on the current density. At higher current densities (1–20 mA/cm^2), where pore formation was supposed to be negligible, aluminum was found[53] to exhibit very similar behavior to tantalum, that is, the Xe^{125} atoms became deeply buried in the oxide.

Davies and Domeij[54] have developed a similar method using α-spectroscopy, with Rn^{222} as the tracer atom. This method was claimed to have the advantage over β-spectroscopy, in that the distribution of tracer atoms which are more deeply buried in a solid may be measured accurately. In the anodization of tantalum and aluminum, the latter being anodized at 20 mA/cm^2, the Rn^{222} tracer atoms became buried at about one third of the way into the oxide from the air/oxide interface, suggesting that the mechanisms of oxidation of the two metals are rather similar. The results for tantalum agreed to within experimental error with those of the previous work[50,53] using Xe^{125}.

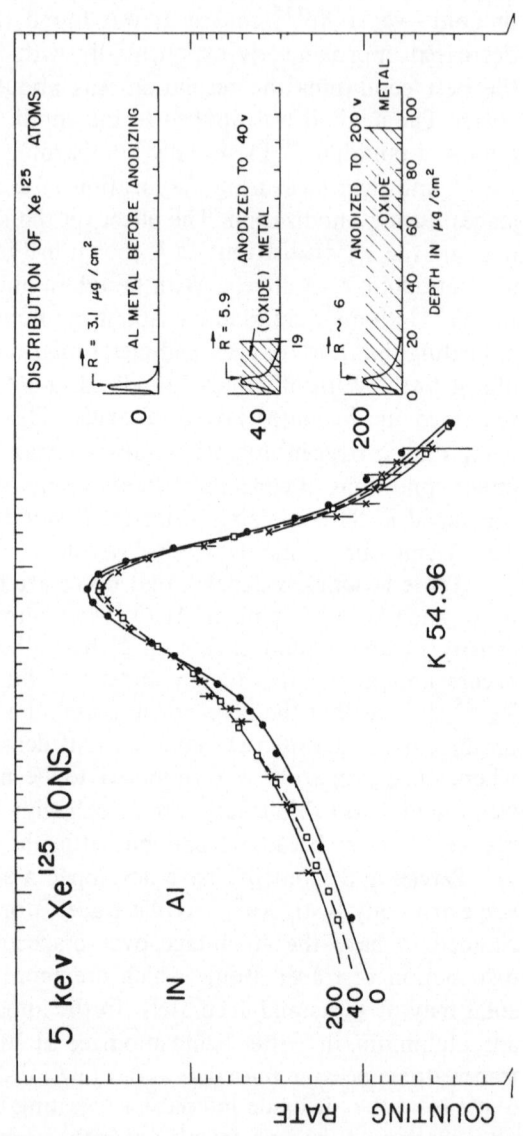

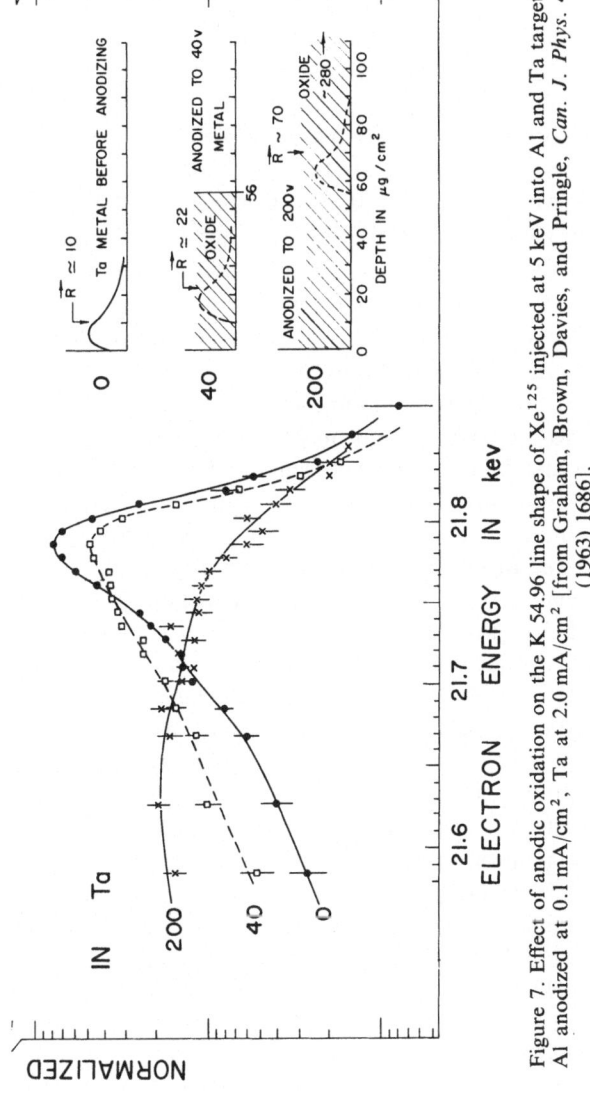

Figure 7. Effect of anodic oxidation on the K 54.96 line shape of Xe^{125} injected at 5 keV into Al and Ta targets. Al anodized at 0.1 mA/cm^2, Ta at 2.0 mA/cm^2 [from Graham, Brown, Davies, and Pringle, *Can. J. Phys.* **41** (1963) 1686].

A further, more detailed, radiochemical investigation of growth of anodic oxide films on tantalum and aluminum has recently been made by Davies *et al.*[55] Two types of experiments were performed. First, "transport number" experiments using Xe^{125} markers as described previously, and second, experiments with neutron-activated metals to determine the amount of metal lost to the solution during the anodizing process. In the former, trace amounts of Xe^{125} were incorporated into a thin surface layer of anodic oxide. The specimen was then anodized, the depth of the embedded tracers being measured at intervals with a β-spectrometer. Metal foils identical to those used in the transport number experiments were neutron activated, and then anodized at constant current. The amount of activity in the anodizing solution was then determined by comparison with solutions in which a known quantity of activated material had been dissolved.

For niobium, tantalum, and tungsten, it was found that no Xe^{125} was lost to the solution and only trace amounts (0.1% Ta, 0.5% W) of the metals were detected. This information could not be obtained for niobium as no suitable radioisotope was available. The transport number t_m for all these metals was about 0.30 ± 0.03, the error being due to uncertainty in the Xe^{125} depth.

The fractional burying for zirconium and hafnium was found to be very small. A considerable amount of Xe^{125} was found in the solution together with a small amount of metal, which was nearly enough to account for the Xe^{125} loss. Oxide thicknesses could not be calculated precisely, but in this case, t_m is quoted as certainly not greater than 0.05, indicating that the oxide growth occurs predominantly by oxygen migration.

For aluminum, two cases were found. When anodization was carried out in a solution of 50 g/liter sodium tetraborate in 95% ethylene glycol–5% water, t_m was found to be 0.58 ± 0.04. However when the aluminum was anodized in aqueous ammonium citrate, t_m varied substantially with current density from about 0.37 to 0.1 mA/cm^2 to 0.72 at 10 mA/cm^2. The experimental error in each value was about 10%, but an error in calculation of the total oxide thickness would have meant that the high current-density value would have been nearer to that obtained in the ethylene glycol solution.

The validity of these results depends upon the assumption that the tracer species is immobile. The absence of diffusion in the static oxides has been established[56] for Xe^{125}, but not in the growing oxide. However it has been shown[54] that in similar experiments with Rn^{222}, no appreciable broadening of the distribution occurred. Some confirmation of these results has come from experiments on tantalum by Cheseldine,[57] who formed an oxide film on tantalum first in formic acid and then in sulfuric acid. He concluded that both oxygen and metal were mobile, and found a value of 0.48 for t_m, which is higher than the value of Davies *et al.*

In conclusion, these experimental results suggest that metal and oxygen are both mobile to comparable extents over a wide range of ionic currents.

(vi) *Consequences of Both Ions Being Mobile*

The tracer work of Davies, Pringle, and their collaborators, as discussed above, indicates that with tantalum, both ions are not only mobile, but the metal and oxygen ions also travel right through the film (in opposite directions) with oxide formed simultaneously at both interfaces. The chief alternative would be that oxide was formed by a sort of recombination process in the interior of the oxide, so that there would be two currents decaying into the oxide from either side, the sum at any depth being constant. This case agrees with the homogeneity of the oxide and the independence of the field on thickness, the alternative case obviously leads to metal excess in one part of the film and to oxygen excess (or lower metal excess) in the other.

The observation that both ions are mobile during the ordinary growth process to comparable extents over a wide range of currents, presents considerable difficulties to the theory.

The chief one is that any model with independent metal and oxygen ion currents would be expected to display a rapid change with field in the ratio of the two currents unless one had a fortu itously close similarity between values of activation energies and activation distances, which is very unlikely to occur with more than one metal.

A first point to consider is whether electroneutrality applies. If it does, the concentration of the more mobile ions (the metal ions) will be controlled, according to the conventional theories, by the concentration of the less mobile ions. As was pointed out in a previous discussion,[60] this model would then give transients similar in some respects to those observed. This model was rejected at the time as an explanation for the transients, because in experiments in which the current is suddenly increased or decreased, the adjustment of the concentration of ions would appear to have to start at the oxide/solution interface and travel through the oxide, thus giving a dependence on thickness which Dewald's work had shown was not present. This objection still seems valid. If one assumes that electroneutrality does not hold and that both currents are controlled by injection the above problem remains, it becomes, in addition, very difficult to explain the transients except in terms of changes at the metal oxide interface, such as unwinding of growth spirals, or other processes which could cause a variation of the disorganization of the metal surface, and hence a variation in the concentration of ions available to enter the oxide. This is a particularly unsatisfactory model, since one has no indication of how the model could give results of the observed type.

The best place to search for an explanation of the observation that both ions are mobile to comparable extents over a wide range of current densities seems to be in some form of linkage between the two currents. The idea proposed in an unpublished discussion by Vermilyea, is that the unsticking of a metal ion tends to release also an oxygen ion. One may think of a bond being broken, or of the removal of a metal ion leaving a local unbalance of charge, which renders the site less attractive to the oxygen ion.

(vii) Dielectric Properties

In our present discussion the importance of the dielectric properties lies in the evidence they give on the structure of the films in relation to the growth process.

These films normally show a tan δ of the order of 0.5 to 1 %, nearly independent of frequency in the audio range, but dependent on formation conditions and annealing (for references see Ref. 1). This implies a wide range of relaxation times. Observations of so-called anomalous charging currents, and of the build up of voltage across films which have been first charged, then discharged for a short time, and finally placed on open circuit, show that this range of relaxation times extends to hours. The nature of the processes giving rise to this spectrum of relaxation times is not generally agreed upon. The choice is perhaps between losses due to the movement of tantalum ions between adjacent sites, and losses due to electronic processes. This has been discussed elsewhere at length.[1] The correlation between the time dependence of the capacity on ceasing formation and the fall in the immediate ionic conductivity suggests that in either case the variations in the losses are associated with the redistribution of tantalum ions among available sites. The explanation in terms of the movement of ions seems the more likely to us, though Smyth, Shirn, and Tripp[61,62] are inclined to favor a variation of electronic conductivity through the thickness of the film, by analogy with the theory which they propose for films annealed in air or vacuum at 200°C and above. An exponential variation of conductivity through the film was shown by Young[1] some years ago to give a constant tan δ. The dielectric properties seem likely to be an important source of information on the structure of the films.

Some measurements have recently been made by Pavlovic[63] on bulk crystalline aggregates of Ta_2O_5 in α and β crystalline forms. The dielectric constant found for the β form was close to the range observed with the anodic films. The α form was much lossier and had a higher dielectric constant. Pavlovic suggested that the α-Ta_2O_5 was more similar to the anodic oxide material on the grounds that the α-Ta_2O_5 has a more open structure, analogous with the oxide glass model of the films. Smyth, Shirn, and Tripp[61] consider, on the basis of Kofstad's work,[64] that such aggregates would be nonstoichiometric and inhomogeneous.

(viii) Oxide of Unusual Properties

Exposure to UV radiation causes the films to grow at field strengths which would otherwise be too small to produce appreciable growth (see Ref. 1). Stimulated growth also occurs at higher fields, and the oxide produced is more normal in properties, that produced at low fields being unusual in hardly contributing to the reciprocal capacitance, as well as in being an abnormally good ionic conductor. When anodized tantalum is heated to 200°C or more, as was first shown by Vermilyea[13] and later investigated in detail by Smyth, Shirn, and Tripp[61,62] an increase of capacitance occurs but with no change in interference color evident to the eye. These effects are of technical importance since the oxide is heated to

250°C or so during the production of "solid" electrolytic capacitors, in which manganese dioxide produced by the pyrolysis of manganous nitrate replaces the electrolyte. It appears that the changes are due to loss of oxygen to the metal. The production of oxygen ion vacancies contributes to the electronic conductivity. If the heating is performed in vacuum, the entire film becomes conductive in capacitance measurements. In oxygen a steady state is set up in which oxygen enters the oxide from the gas phase at the rate at which it passes through into the metal. The film may be replenished with oxygen by reanodizing. An exaggerated form of the effect described in section III(iii) occurs for films formed to low current density or annealed at only 100°C. A specific example quoted by Smyth *et al.* was a specimen formed to 75 V, heated at 400°C for 30 min in air, and reanodized with 85 V applied. The current remained below 0.1 A/cm^2 for about an hour before starting to rise, and only then did a color change commence, indicating an increase in thickness. The temperature dependence of capacitance of such annealed films is such that $1/C$ increases first rapidly with decreasing T and then becomes almost constant (Figure 8). This was thought by Smyth *et al.* to indicate that the oxide was inhomogeneous through its thickness and had a varying electronic conductivity. As the conductivity falls with decreasing temperature, the effective thickness of the oxide in capacitance measurements increases till it is equal to the geometric thickness. It is known that oxide deficient in oxygen conducts electronically. The lack of contribution to the reactance of the photogrown oxide could also be explained by electronic conductivity. However, with this oxide the ionic conductivity is high not low. That changes in the film precede photogrowth at low fields is shown by the fact that photogrowth does not start immediately when the UV radiation is applied. The electronic photocurrent gradually decays and the "dark" current, defined as the current observed when the irradiation is momentarily cut off, increases slowly. This increased dark current seems to correspond to the photoinduced growth current.

The occurrence of these two types of unusual oxide is of importance to our topic because it clearly provides a clue to the nature of the ionic transport process, but the significance of this clue is not presently understood. Oxide of unusual properties is also formed when solutions with a low activity of water are used in the anodization process.

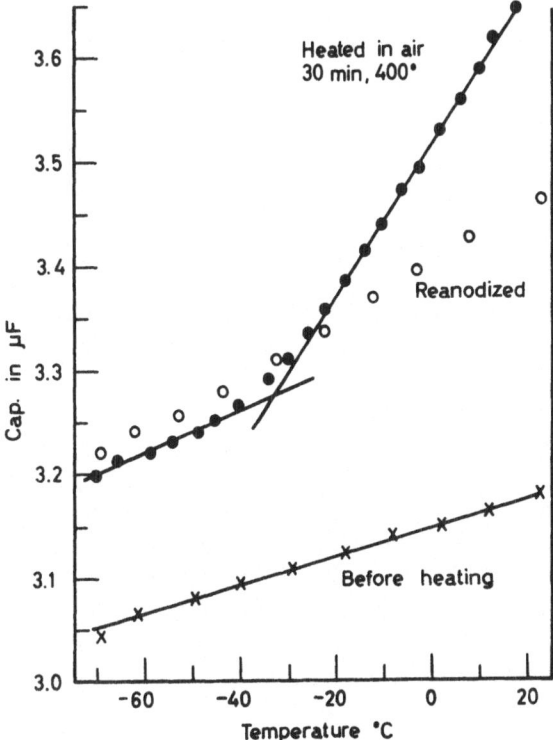

Figure 8. Temperature dependence of capacitance of un-
heated, heated, and reanodized Ta–Ta$_2$O$_5$ [from Smyth,
Shirn, and Tripp, *J. Electrochem. Soc.* **110** (1963) 1264].

(ix) *Adhesion to Substrate*

Another rather extraordinary effect which is unexplained, but
which must be of significance to the mechanism of growth, is the
remarkable disappearance of the usually very strong adhesion of
the oxide to the metal occurring with tantalum and to a lesser
degree with niobium when the film is formed on chemically polished
tantalum[65] or, according to recent reports, in solutions containing
fluoride ions. Excellent adhesion is obtained with chemically
polished metal if it is leached in boiling water before anodizing.
Evidently some modification of the metal/oxide interface occurs:
possibly the retention of a layer of oxyfluoride, or possibly an

accumulation of vacancies. This layer of different ε or ionic conductivity may be detected by capacitance measurements.[1]

(x) Mosaic Structure Theory

Dignam[66,67] has treated a model of the oxide which he claimed to account for the chief features of the experimental results. The oxide was supposed to be built up of microcrystals—small volumes of molecular groupings approximating to crystalline oxide. The rate of jumping of tantalum ions between the units was supposed to limit the current and to be determined by the effective field, a cavity field, and hence by the dielectric polarization of the oxide.*

(xi) Reactions at the Interface

At higher temperatures, as Smyth et al. have pointed out, the boundary between the metal and the oxide is likely to be blurred because of thermally activated movement of oxygen into the metal. However, this effect is probably negligible below 100°C. It would be detectable as a change in the lattice constant of the metal. The macroscopic electric field will doubtless penetrate into the metal by an Å or so. This, and the local fields of the ions, will cause reaction of oxygen ions arriving at the interface. Some disorder will occur because of the departure, one by one, of the metal ions into the oxide.

At the oxide/solution interface, one may visualize the reaction as occurring by tantalum ions popping out and capturing O'' from species in the solution, or suppose that O'' ions enter the oxide and the reaction occurs in some sort of transition layer in the oxide surface.

Vermilyea[68,69] has reported some studies of processes involving proton exchange between very thin films and the solution. Capacitance and pH vs. potential measurements were made. The potential was found to vary by about 0.057 V per pH unit except with solutions containing Li^+ and Na^+ with pH > 4. The proton exchange gave rise to large absorption capacitance. A space charge in the oxide due to protons was believed to respond to the AC signal and to depend on DC bias. Since these effects are not too directly concerned with our main topic, the reader is referred to the originals for details.

*A full account has now appeared, J. Electrochem. Soc. 112 (1965) 722.

(xii) Experimental Techniques for Studies of Ionic Currents

Experimental methods of investigating the ionic conduction process must start from some method of determining the thickness of the oxide, since the thickness must be known to determine the field. The rate of change of thickness determines the ionic current unless variations of composition occur, or ions of unusual valency are produced which result in chemical reactions at the interface in which hydrogen liberation accompanies oxide production. Most of the available methods are insufficiently reliable for satisfactory kinetic studies. Only the optical methods seem to be reasonably unequivocal. An optical thickness nD is usually found. If the oxide is shown to be uniform, a determination of the refractive index n gives D. Some optical techniques of high accuracy have been described elsewhere in detail.[1] The technique of ellipsometry has recently been explored and some discussion will be given here.

The ellipsometric method has the special advantages that measurement may be made with the oxide film still immersed in the forming electrolyte, and that very thin films may be measured. In the usual descriptions of the method the optical constants of the substrate, which are required for determination of the thickness, are determined by measuring the bare surface. In practice it is not easy to obtain a bare metal surface without recourse to ultra-high vacuum techniques, and other methods must be used to obtain the optical constants of the substrate. The ellipsometric technique gives[70] the relative changes in phase Δ and amplitude $\tan \psi$, of the p and s light components (electric vectors parallel and perpendicular to the plane of incidence, respectively) on reflection from the specimen. These two quantities are plotted to give a graph which may then be fitted by trial and error to theoretical curves.[71] Once this has been done, it is a simple matter to read off values for the thickness, and to find the ionic current by differentiating with respect to time.

It would be straightforward to devise a self-balancing attachment for the ellipsometer which would be fast enough to take measurements continuously at reasonable current densities, thus avoiding the interruption of the film growth, and, therefore, avoiding transient conditions. This has not yet been done.

The theory, construction, and use of the ellipsometer have recently been reviewed by Zaininger and Revesz.[72] The technique has been applied to oxide films on silicon,[71,73,74] tantalum[21] and titanium[75] among others, adsorption-desorption processes of organic compounds,[76,77] and anodic formation of calomel on mercury electrodes.[78]

Recent work by Kumagai and Young[21] on tantalum has shown the importance of not neglecting the "pre-existing" oxide film when determining the optical constants of the substrate, particularly with reactive metals like tantalum. Tantalum, which had been freshly etched in HF gave apparent (wrong) values of n_2 and k_2 (for the metal) of about 2.4 and 2.3, respectively, for an angle of incidence ϕ of 70° and $\lambda = 5461$ Å. The change of ψ and Δ with time is shown in Figure 9, and whereas the rate of change is slow, this is because most of the pre-existing film had grown previous to the first measurement. (Archer's method[71] for silicon depends on its slower oxidation rate in air, and on the nature of the

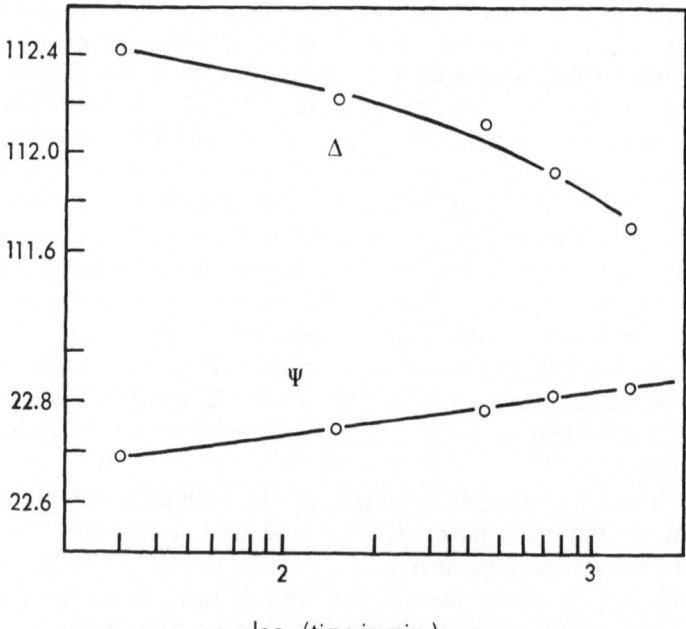

$\log_{10}$ (time in min.)

Figure 9. Time dependence of optical behavior of unanodized tantalum [from Kumagai and Young, *J. Electrochem. Soc.* **111** (1964) 1411].

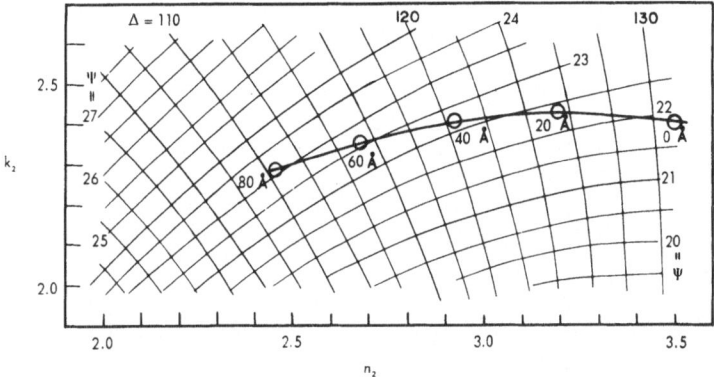

Figure 10. The grid shows ψ and Δ calculated for bare metal with optical constants shown on the axes. The extra line shows the dependence of ψ and Δ on thickness of oxide for $n_2 \simeq 3.5$ and $k_2 \simeq 2.4$. If these films were neglected the value of n_2 would be wrongly estimated, for example as 2.4 for ~ 80 Å of oxide [from Kumagai and Young, *J. Electrochem. Soc.* **111** (1964) 1411].

values of n_2 and k_2 for silicon.) Figure 10 shows the effect on n_2 and k_2 of wrongly assuming that values of ψ and Δ which would be observed with tantalum with a series of very thin oxide films present were for the bare metal. Values for the optical constants of the metal and oxide were obtained by curve fitting. They were in satisfactory agreement with those previously obtained by Masing, Orme, and Young[15] using a method, new in this field, which involved, in particular, measuring the ratio of maximum to minimum reflectivity in p light as the thickness was increased. The value of the index of the oxide agreed with an earlier value obtained by the Becke method. Figure 11 shows data for tantalum which indicate no change in optical properties over a wide range of thickness.

(xiii) Anodic Oxide Films on Semiconductors

The growth of anodic oxide films on semiconductors is of topical interest in microelectronics. The anodic production of oxide has the advantage over thermal methods in that it may be performed at room temperature, thus avoiding unwanted alloying and doping which may occur at high temperatures. Kinetic studies are complicated by several practical problems, for example, cutting and preparing the surface of the often brittle specimens and making an

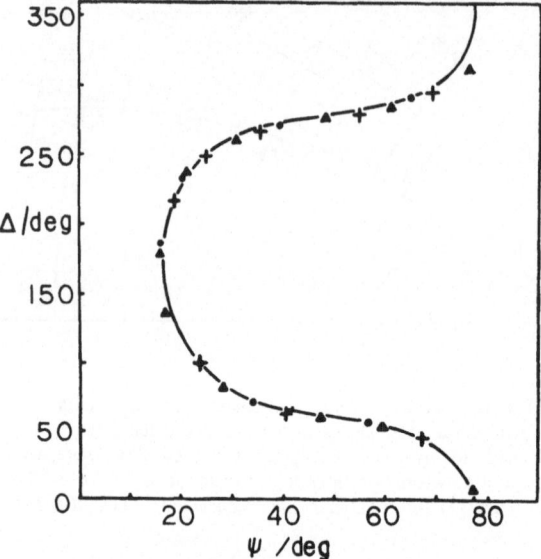

Figure 11. Unpublished ellipsometric data obtained by Zobel and Young for tantalum anodized at constant voltage in 0.2 N H_2SO_4, measured *in situ*, $\phi = 67.5°$. As the oxide thickness increases, the points should lie on the same curve, provided the optical constants of the oxide do not change. Three cycles of the curve are shown here (first—dots, second—triangles, third—crosses). The line is calculated using the values of the optical constants of the metal obtained by Kumagai and Young,[21] $n_{oxide} = 2.26$ and $n_{soln} = 1.334$.

ohmic electrical contact. The high resistance of some semiconductors can lead to an appreciable voltage gradient along the specimen.

(*a*) *Silicon*—The fact that silicon can be anodized was first reported by Güntherschulze and Betz.[7] Schmidt and Michel[79] investigated a number of aqueous electrolytes, and found that the best films were obtained with 0.04 N KNO_3 in N-methylacetamide which has a dielectric constant of 179. This solvent was superior to N-methylformamide. Thickness measurements, using weighing and interference colors, indicated field strengths of about 2 or 3 × 10^7 V/cm with current efficiencies of about 1 or 2 %. These very high fields make silicon of special interest for academic studies.

Politychi and Fuchs[80] stripped anodically formed films from silicon, and studied them with an electron microscope. They found that films formed in aqueous electrolytes tended to be porous (a possible exception being those formed in caustic soda). Films formed in N-methylacetamide were reported not be be porous, in contradiction with more recent observations.[74]

Investigation of the electrode reactions involving solution species taking place during anodization of silicon have been made by Duffek, Mylroie, and Benjamini[81] for low resistivity material in several electrolyte solutions in N-methylacetamide. The rate of anodization depended on the water content of the N-methylacetamide, and was followed by determination of the current efficiency. It was postulated that water (which is formed as an electrolysis product) is in fact necessary for the anodization process. A similar study has been made using ethylene glycol solutions.[82] Rabinovitch and Borrer[83] measured the rates of oxidation of 1–3 ohm cm p-type silicon in sulfuric acid and caustic soda solutions over a range of temperatures. No experimental results were given, but they stated that the voltage thickness curve was "convex," indicating a decreasing density as the film thickened. However, as the current densities used were rather high (up to 800 mA/cm^2) it is probable that their films had suffered dielectric breakdown in most cases.

Determinations of the refractive index and thickness of oxide films formed by various methods on silicon have been made by Archer[71] and by Claussen and Flower[74] using ellipsometry. Archer found a refractive index of 1.362 for an oxide film formed in an aqueous solution of 0.1 M sodium borate and 0.1 M boric acid. Claussen and Flower found that the refractive index of the oxide formed in 0.0025 N KNO$_3$ in anhydrous N-methylacetamide was 1.468 and that of films formed in 0.04 N KNO$_3$ in N-methylacetamide with 6–8% water was 1.417. Zobel and Young using Lewis' method[84] with liquids of known refractive index, and ellipsometry, have observed a refractive index for films formed in the above aqueous borate solution of about 1.4. Care has to be taken to check that film breakdown effects do not render the film untypical.

The high resistance films formed in N-methylacetamide solutions have been used in experimental sintered anode silicon electrolytic capacitors.[85] Thermally oxidized films have been anodized in the same solution to improve their leakage resistance for use in capacitors.[86] Anodic oxide films on silicon may be used as sources for doping elements in integrated silicon devices.[87] The dopant is incorporated into the oxide from the electrolyte during anodization.

Waring and Benjamini[88] found that at constant current density the luminescence during anodization of silicon increased as some low power (between 1 and 3 depending on the resistivity and type of the silicon) of the "net voltage."

The anodic oxidation of silicon in pure water has been reported[89] to give highly insulating oxide films. The "electrical strength" of these films was quoted as 5 to 16×10^6 V/cm.

(b) *Germanium*—Anodic oxide films cannot usually be grown on germanium in aqueous acid or alkaline solutions, since the oxide formed is soluble. Uniform oxide films on germanium may be grown in nonaqueous solutions. Schmidt and Michel[79] reported that both n- and p-type Ge could be anodized in the same solution used for silicon, i.e., a dilute solution of potassium nitrate in N-methylacetamide. Zwerdling and Sheff[90] grew valve-metal-type oxide films in an anhydrous solution of sodium acetate in glacial acetic acid up to a thickness of 1240 Å on both n- and p-type germanium. The films were shown to be GeO_2 by their infrared absorption in the region 2 to 20 μ. The amount of GeO_2 formed was determined by a chemical method after dissolving the films in dilute aqueous caustic soda, and from this, the ionic current and field strengths were estimated. These latter were in the range 1 to 4×10^6 V/cm[1] for 10 to 30 A/cm^2, and current efficiencies were 10 to 20%. With n-type germanium at high current densities the anode voltage increased to abnormally high values, then decreased. The voltage–time curves at constant current showed two or three linear portions. The nonideal behavior prevented further study of the kinetics. There was found to be no difference between the crystal faces within the experimental error.

Wales[91] anodized intrinsic polycrystalline germanium in a solution containing approximately $0.6 \times 10^{-3} M$ LiNO$_3$, 0.1 M acetic acid, 0.25 M H$_2$O and $10^{-5} M$ GeO$_2$. Films were grown to a thickness of about 7000 Å, with a differential field strength of 2×10^6 V/cm. The breakdown voltage at 100 μA/cm^2 was 180 V, and at 250 μA/cm^2 was 150 V. The current densities used ranged from 25 to 250 μA/cm^2 and the current efficiency was about 70 to

80 %. Effects of electrolyte composition on the rate of growth were studied and the above optimum concentrations were found to give the most uniform characteristics. The concentration of water had a profound effect on the growth rate, probably because of dissolution of the oxide film. The effects reported on ageing of the solutions may also have been due to hydrolysis of the GeO_2. Film thicknesses were found, as previously,[90] from chemical analysis of the amount of germanium present in the films, and the pre-existing film thickness was assumed to be zero. The oxide was shown to be GeO_2 by X-ray analysis. Interference colors in the films were tabulated and compared with Vasicek's results[92] for GeO_2 films on glass.

In a later paper, Wales[93] measured capacitance and dissipation factors for anodic GeO_2 films formed on germanium in the manner described above. Contact to the dry oxide film was made by means of a small drop of mercury. Capacitances were measured both in solution and dry, lower values being obtained in solution for the thinner films. This was explained in terms of porosity of the thinner films, but could also be due to a contribution from the oxide–electrolyte interface (see Vermilyea,[68,69] section III.2.(xi)) which would be more significant at small oxide thicknesses. The dielectric constant of the GeO_2 was calculated to be 6.4 at 1 kcps. The capacity decreased about 6 % with 100-fold increase in frequency for the dry films. The dissipation factor was nearly independent of thickness, but varied with frequency, increasing toward higher and lower frequencies from a minimum value of about 0.04 at 1 kcps. The DC resistivity of the dry films was about 3×10^{12} ohm cm which is somewhat higher than the value of 2×10^{10} ohm cm quoted previously for wet films.[90,94]

(c) Indium antimonide—As first shown by Dewald in 1957[95] InSb is of special interest because of a marked dependence of behavior on crystal face. The crystal faces of InSb show markedly different surface structures and were found to show different oxidation rates at low fields; in particular, the (111) and (332) faces oxidized at 10 times the rate of the (110), and $(\overline{111})$ and $(\overline{332})$ faces, but at high field strengths all faces oxidized at the same rate. This implied that the kinetics at low fields are controlled by the oxide–substrate interface and not by the bulk oxide. This was particularly surprising because the transient behavior was very like that of tantalum. The

composition of the films was shown to be almost exactly 1:1, In:Sb, but at the oxide–electrolyte interface, the concentration of Sb fell sharply to zero. Since antimony oxides are soluble in aqueous solutions of KOH, an interstitial antimony ion arriving at the oxide surface should dissolve and not form oxide so that the simple model of mobile cations in a stationary oxygen lattice is not obeyed for this system.

The Tafel plots (log formation rate *vs.* relative field) were not linear, and it was suggested that they would be made up of two linear portions, the slope being higher at higher field strengths. The work of Venables and Broudy[96] confirmed by Harkness and Young[97] using optical and capacitance measurement techniques, indicated no difference in kinetics between crystal faces below a "transition voltage" which was different for different crystal faces, and which increased with increasing current density. Above this transition voltage, the film reached very different thicknesses on different faces.

Photo-induced growth with InSb has been investigated by Venables and Broudy[96] and by Mueller and Jacobson.[98]

IV. ELECTRONIC CURRENT

1. Introduction

A discussion will be given of electronic currents through sandwich structures of the type: tantalum (or other substrate metal)/oxide film/metal counterelectrode. The thickness of the oxide film has varied from 25 to 5000 Å. The counterelectrode has usually been deposited on the oxide by evaporation, but pressure contacts, mercury droplets, and electroless plating[1] have also been used. The behavior of the system metal/oxide/electrolytic solution is more difficult to interpret and little can be added to a previous article.[1] Even with the simple metal/insulator/metal system there is disagreement about which mechanisms control the current under the various conditions of temperature, thickness, and field. However, recent work has clarified the picture with regard to the choice of mechanisms, and experimental results are beginning to accumulate. Some effects, such as the negative resistance, which has been observed with films which have been subjected to a preliminary breakdown, can be explained only very tentatively.

We first review, before giving a discussion of the experimental findings, what is to be expected for a macroscopic metal/insulator/metal sandwich, and what differences are expected when the insulator is made thinner.

(i) Electronic Conduction in Amorphous Oxide

Some special features as contrasted with those of ordinary crystalline semiconductors are to be expected in the electronic conductivity properties of the present oxides because of their amorphous nature.[99,100] It has often been assumed that these amorphous oxides will have a conduction band and valence band just as for crystalline materials. However, as with the conduction associated with a disordered array of impurity centers (Mott and Twose[101]), "it is unlikely that the phases of electrons scattered by a completely disordered lattice will match sufficiently well for electrons of any energy to show a negative effective mass." Hence, one should perhaps not expect to observe conduction by positive holes. (Empty states in the top of the valence band of crystals behave like positively charged, positive mass electrons because they are, so to speak, the negative of negatively charged electrons with negative effective mass.) The model treated by Mott and others (reviewed by Mott and Twose[101]) of a random array of impurity centers in a crystalline lattice is probably quite applicable to the present systems. Although an extended discussion of the peculiarities of impurity conduction in relation to the present systems is not available, the concept of impurity conduction has been applied by several authors, in particular by Mead and Hickmott (see later sections). The electrons move "from one positively charged donor to another, the host crystal acting as a dielectric medium in which the random lattice is embedded." The characteristic wave functions are either localized, or, in the case of a concentrated array under certain conditions, delocalized. In the first case, conduction will ordinarily be by thermally activated hopping from one site to the next. In the second case, a type of metallic conduction can arise.

(ii) Potential Distribution Within the Film

The first step is to consider the energy-band diagram of the system[102] (treating the oxide as if it were a crystalline semiconductor). Two assumptions are usually made to obtain the band diagram of metal–semiconductor contacts. First, one has, of course, a constant Fermi energy through the system at equilibrium. Second, the energy bands at the contact are supposed to be displaced, with respect to their position relative to the Fermi level in the bulk semiconductor, in such a way as to equalize the vacuum levels "just outside" the metal and semiconductor (Figure 12). The work function ϕ of the metal is defined as the difference between the vacuum level and the Fermi level of the metal. The electron affinity χ of the semiconductor is defined as the difference between the vacuum level and the bottom of the conduction band. Hence this second assumption places the conduction band at the contact $(\phi - \chi)$ above the Fermi level. This separation between conduction band and Fermi level is in general not the same as that within the bulk semiconductor. The second assumption is actually an oversimplification. The charge configuration at the interface is in general quite uncertain as there can be a dipole sheet or electrical double layer, usually discussed in terms of "surface states," which can give an extra potential drop across the interface. We use $(\phi - \chi)$ as a shorthand, with these reservations implied. It is clear that

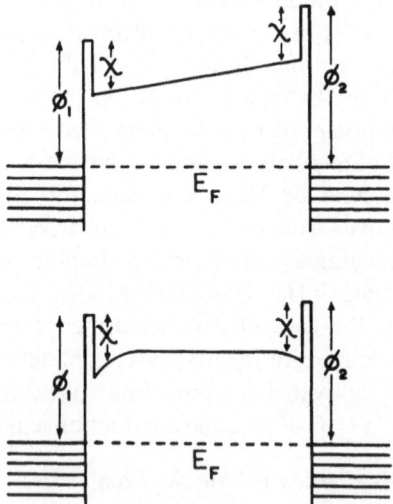

Figure 12. Band diagram for semiconductor with two metallic contacts. Top, very thin film; bottom, thick film.

the work function of the clean metal may in practice have little bearing on the behavior of actual metal semiconductor contacts. In equilibrium the concentration of electrons in the conduction band just inside the semiconductor[103] is

$$N_0 = N_c \exp\left[-\frac{(E_c - E_F)}{kT}\right] = N_c \exp\left[-\frac{(\phi - \chi)}{kT}\right]$$

where N_c is the usual effective density of states with the bottom of the conduction band at E_c, E_F is the Fermi level and we assume $\phi - \chi \gg kT$. The variation with distance of the electrostatic potential in the semiconductor in the region from near the interface to the electrically neutral bulk region is given by the Poisson–Boltzmann (or Poisson–Fermi) equation in terms of space charge due to electrons, holes, and ionized donor or acceptor levels. The surface potential, that is, the total change in potential into the bulk is given by the negative of the difference between the above separation between E_c and E_F at the surface and the separation value determined by electroneutrality in the bulk. The bands bend up or down toward the surface, making the surface layers more p (less n) or more n (less p) type, respectively, than the bulk material. There is a considerable literature on such effects which are of great technological importance.[167]

Since the space charge in the semiconductor is limited in density, the field will tend to become constant through the thickness with a sufficiently thin film (Figure 12). It will be given by

$$E = \left[\frac{(\phi_1 - \chi) - (\phi_2 - \chi)}{d}\right]$$

where d is the thickness. This field can be as large as 0.5 V/100 Å $= 0.5 \times 10^6$ V/cm.[104] The concentration of electrons in equilibrium will then be exponentially dependent on the distance

$$n = N_0 \exp\left[\frac{e(\psi - \psi_s)}{kT}\right] = N_0 \exp\left(-\frac{eEx}{kT}\right)$$

provided the film is thick enough to apply these equations. Mott and Gurney[103] showed that in the case of an insulator, where the concentration n_0 of electrons in the conduction band in the bulk is very small, and $N_0 \gg n_0$, the Poisson–Boltzmann equation takes a particularly simple form and the potential varies with distance into the oxide in the region where $n \gg n_0$ according to

$$\psi - \psi_s = -\left(\frac{2kT}{e}\right) \log\left(\frac{1 + x}{x_0}\right)$$

where $x_0 = (2\pi N_0 e^2/\varepsilon kT)^{-\frac{1}{2}}$ (electrostatic cgs as in original). The thickness of a "very thin" film is therefore of the order x_0. To quote the original values of Mott and Gurney

$(\phi - \chi)/1$ eV	0.1	0.5	1.0
N_0/cm^{-3}	10^{17}	10^{10}	10^2
x_0/cm	10^{-6}	10^{-2}	10^2

(iii) Ohmic Region

Both the macroscopic and the thin-film systems are to be expected to give Ohm's law for very small applied voltages ($\Delta\varphi/e \ll kT/e$). In practice there is little data for this region because the currents are too small to measure conveniently. "Anomalous charging currents" flow for some time at zero voltage if fields have previously been applied. The charging time associated with the capacitances of cables can give similar but spurious effects. For the limiting case with thick films and bulk control, the ratio dV/dI is simply given by the bulk conductivity. The very thin film case, with a constant built-in field due to the contact potential difference, can be formulated easily enough in the usual way, but the equations are of little experimental significance. With extremely thin films the tunneling effect between the metal electrodes gives an Ohmic characteristic at low fields with relatively large currents.

Ohm's law applies quite generally for small voltages. If the current is expanded in a Taylor series in terms of the applied voltage the first nonvanishing coefficient must be for an odd power of $\delta\varphi$, since the current changes sign with φ, and in general the coefficient of the first power of $\delta\varphi$ will be nonzero.

(iv) Space Charge Limited Current

As the field is increased, considering first the macroscopic case with $N_0 \gg n_0$ ("Ohmic contact–virtual cathode"), the diffusion component $eD\, \partial n/\partial x$ of the current ($I = ne\mu E + eD\, \partial n/\partial x$) eventually becomes negligible in comparison with the migration term. Assuming that no traps are present and that the space charge is due solely to the electrons swept out of the virtual cathode, the space charge limited current law of Mott and Gurney[103] may then be observed (e.g. Ref. 105) under certain conditions: $I = (\frac{9}{32}\pi)\varepsilon\mu V^2/d^3$, where V is the applied voltage. This expression is not explicitly dependent on $(\phi - \chi)$ (i.e., on the electrode materials and hence on which way the current flows), but it involves the assumption that

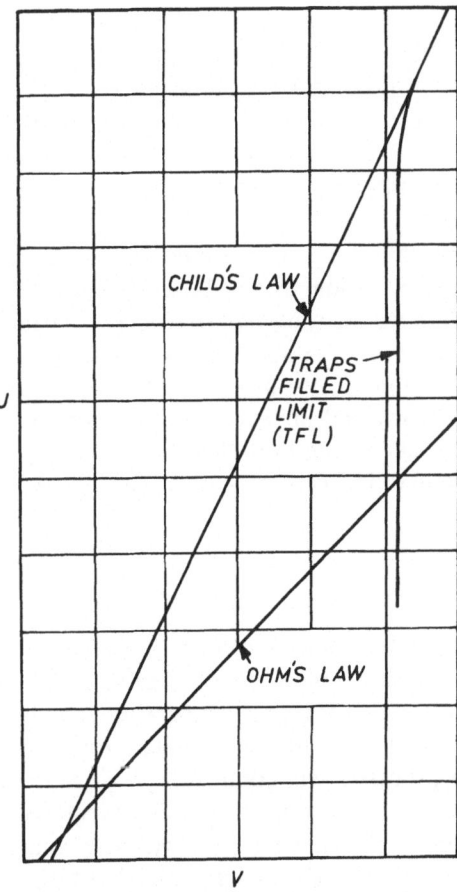

Figure 13. Log–log plot of limiting current density (J) *vs.* voltage (V) characteristics for space-charge-limited currents in an insulator with traps [from Lampert, *Phys. Rev.* **103** (1956) 1648].

the stock of electrons N_0 just inside the semiconductor at the negative electrode is large enough and that it is replenished without appreciable depletion by thermionic emission from the metal. This implies an assumption about $(\phi - \chi)$ with relation to the conduction properties of the insulator. If traps are present, the current rises very rapidly as the traps are filled with injected carriers, and the space charge limited law is observed when they are full (Figure 13). Eventually, as the applied voltage is further increased, the space charge due to the electrons in transit may become negligible, even with the macroscopic insulator. If this occurs with the virtual cathode still operative, Ohm's law is again expected but with $I = N_0 e\mu E$, as compared with the low field law $I = n_0 e\mu E$ where n_0 and μ are the effective concentration of electrons and their mobility, respectively, in the bulk electrically neutral material.

With very thin films one would expect the space-charge limited-current-law region to be absent, because the total space charge in the film will be too small to change the field appreciably. The question, of course, is how thin is very thin. A considerable literature[105-115] now exists on the subject of space charge limited currents in the many and involved situations which arise with the macroscopic system as one introduces various complications such as traps, recombination centers, and injection of both holes and electrons.

(v) Higher Fields: Summary

As the applied voltage is further increased, the injection of electrons must increase if more current is to flow. This occurs by the Schottky mechanism[116] at higher temperatures and by the Fowler–Nordheim mechanism[117] at low temperatures. With thin films these injection processes will control the I–V relation. With sufficiently thick macroscopic samples the voltage required to set up this injection will be negligible compared with the voltage across the bulk of the insulator. Processes are available which can allow the bulk conductivity to increase rapidly with further increase of field. These include (a) the so-called Poole–Frenkel effect,[118] which is the field assisted thermally activated freeing of electrons from traps (analogous to the Schottky effect), (b) field assisted tunneling from traps [analogous to the Fowler–Nordheim mechanism[117]], (c) the Zener effect,[117] and (d) avalanche effects.[119]

(vi) Injection Processes

The Richardson law gives the exchange current at equilibrium between the metal and the conduction band of the semiconductor.[120] When the equilibrium concentration of electrons within the semiconductor is depleted, the unbalance provides net current. Saturation at the exchange current does not occur because the applied field increases the rate of injection. At ordinary temperatures this occurs because the field reduces the height of a potential-energy barrier, restricting the flow of electrons from metal to semiconductor. The Schottky formula is obtained by assuming an image-force law for the barrier, that is, the force on the electron is taken as $-e^2/4\pi\varepsilon(2x)^2 + |E|e$ for a field E causing increased emission. This gives in effect a modified $(\phi - \chi)$ in the Richardson equation. At lower temperatures the thermal activation over the barrier becomes negligible. Tunneling through the barrier at very high fields takes over. The W.K.B. approximation with various forms of potential-energy barrier gives the Fowler–Nordheim law and various improved versions.[117] These involve expressions of the type $I = AE^n \exp(-B\varepsilon^{\frac{3}{2}}/E)$. (This is for $T = 0°K$.)

(vii) Bulk Processes at High Fields: the Poole–Frenkel Process

The Poole–Frenkel effect is the high-field assisted thermally-activated escape from traps. The derivation of an expression for the rate of escape from traps is similar to the derivation of the Schottky equation. Because the charge on the trap is fixed in position, the attractive force is proportional to $1/x^2$ instead of to $1/(2x)^2$. This gives an apparent difference of 2 in the slope of the plot of log I vs. $E^{\frac{1}{2}}$ as compared with the Schottky mechanism.

The usual expression for the Poole–Frenkel effect is thus the rate of escape from traps. This gives the current in the situation where only a small fraction of traps are ionized at any given moment, and where the field in the electrically neutral oxide controls the current. In a thick enough film the ionized traps near the interface will provide space charge which will accumulate until the field across the interface is increased to the value necessary to keep the rate of injection at the required level. With a thin enough film the injection rate should eventually, as the field is increased, control the current. The Poole–Frenkel effect would then only affect the concentration of electrons in transit. The contribution which is made to the current by each electron which breaks free might be taken as independent of the field, since the electron will travel until again trapped, and the distance between traps is independent of field. However, the usual method is to assume that the rate of escape from the traps gives the conductivity. This leads to an expression $I = EG_0 \exp(\beta E^{\frac{1}{2}}/kT)$. If one plots $\log I$ vs. $E^{\frac{1}{2}}$ and equates the slope to β/kT, the value obtained will be appreciably different from that obtained by plotting $\log(I/E)$ vs. $E^{\frac{1}{2}}$, but both methods will probably give linearity. This merely emphasizes that comparisons between the theoretical and experimental values of β must be treated with caution, a point to which we shall later return.

(viii) Tunneling from Traps

The escape of electrons from traps by tunneling[117] bears the same relation to the Poole–Frenkel theory as does the Fowler–Nordheim theory to the Schottky theory. It gives a law similar to the Fowler–Nordheim law.

(ix) Zener and Avalanche Effects

In addition to the Poole–Frenkel effect and the field-induced tunneling from traps to conduction band states, the Zener effect (field-induced transitions from valence band to conduction band) and various forms of avalanche breakdown effects, can give a bulk conductivity rising sharply with field.[119] These effects are difficult to assess in the present systems, because little is known about the electronic states in amorphous oxides, the electronic transport process, or the lattice vibration spectrum.

Avalanche multiplication occurs when the high electric fields can accelerate electrons to high energies at which they free other electrons by impact ionization. The avalanching effect causes the density of electrons to increase with distance into the dielectric. Higher field strengths will be required in thinner samples to give the same current. Because of the statistical nature of avalanching, the prebreakdown currents will be noisy. Collective breakdown occurs when electron/electron interaction predominates over electron/phonon interaction, and a sufficiently high density of free electrons is present initially. The distribution of electron energies is determined by a Maxwellian distribution characterized by a temperature greater than the temperature of the lattice. As the field is increased a critical value is reached at which the free electrons gain energy faster than it can be transferred to the lattice for any temperature of the electron distribution, and a high rate of ionization occurs. This "breakdown field strength" is independent of the thickness of the sample. The process will be considerably modified by a high density of traps. A strong temperature dependence of the breakdown field can occur because the density of trapped electrons decreases exponentially with increasing temperature.

The Zener effect will require very large fields with wide band semiconductors. In p–n junctions in silicon, avalanche rather than Zener breakdown normally occurs except in very thin junctions.[117]

(x) *Extremely Thin Films: Tunneling from Metal to Metal*

With extremely thin films, tunneling from metal to metal may be the dominant mechanism even at high temperatures, and will certainly be so for low temperatures, where thermally activated processes are frozen out. The theory of this mechanism was considered many years ago by Holm and others in connection with the low resistance of electrical contacts where a thin oxide or other tarnish film will usually be present.

The theoretical treatments[121–134] are mostly based on a single electron approximation with some assumed form of potential-energy barriers. Rectangular and trapezoidal barriers and barriers smoothed by "taking into account" image forces (there are images in both metals) have been treated using the semi-classical W.K.B. approximation in which a solution of the form $\exp i\alpha(x)$ is assumed with $\alpha(x)$ varying slowly with distance x, so that the transparency factor is of the form $Ae^{-\beta}$ with β proportional to $\int (V - W)^{\frac{1}{2}} \, dx$ where $V(x)$ is the potential energy, W is the total energy. Free electron bands have mostly been assumed for the metals, either with $T = 0$ or $T \neq 0$. The theoretical treatments are hence of a very oversimplified model so that detailed comparison with experiment is difficult. The periodic structure of the insulator has been allowed for by taking an effective mass for the electron. The situation where the metals are in the superconducting state is obviously very difficult, and Bardeen[126] has considered the problem from a many-particle viewpoint. However, the main features to be expected in the ordinary case are fairly clear. There will be a very sharp, exponential-like fall in conductivity with increasing barrier width. At low voltages the characteristic will, of course, be Ohmic, but a Fowler–Nordheim type of law is expected for high fields. Temperature dependence will be smaller than for thermally activated processes but will be present because of the change in distribution of electrons according to Fermi–Dirac statistics, variations in effective dielectric constant[128] (which will affect the image force), small changes in dimensions due to thermal expansion, and scattering due to phonons.

The difficult question of the transit time of electrons crossing the barrier has been considered by Hartman,[131] who treated the passage of a Gaussian wave packet through a rectangular barrier.

(xi) p–n Junction Effects

It has often been suggested that rectification by anodic oxide films could be due to p–n junction effects.[1] It has been shown by Kofstad[64] that both p- and n-type crystalline Ta_2O_5 may be prepared by appropriate heat treatments at various partial pressures of oxygen. The conductivity type was established by the sign of the thermoelectric e.m.f. and by the dependence of conductivity on partial pressure of oxygen (Figure 14). Sasaki's recent proposal[135,136] that a thin layer of oxide near the metal may be n-type (excess metal), the major portion of the oxide nearly intrinsic, and a thin outer layer rich in oxygen and, hence, p-type, is reasonable. However, the conduction properties of such a system would not be expected to be identical with those of a p–i–n junction in single crystal silicon with the p and n regions truncated by the metal contacts. As already stated, even the validity of the band theory itself for the present amorphous materials is questionable. It is certainly clear that any p–n junction effects will be likely to disappear with thin films.

Rectification is always observed with the "wet" system except where nonaqueous solutions free of protons are used. The forward direction is obtained with the substrate metal negative. Rectification is often observed with the dry system, depending on the

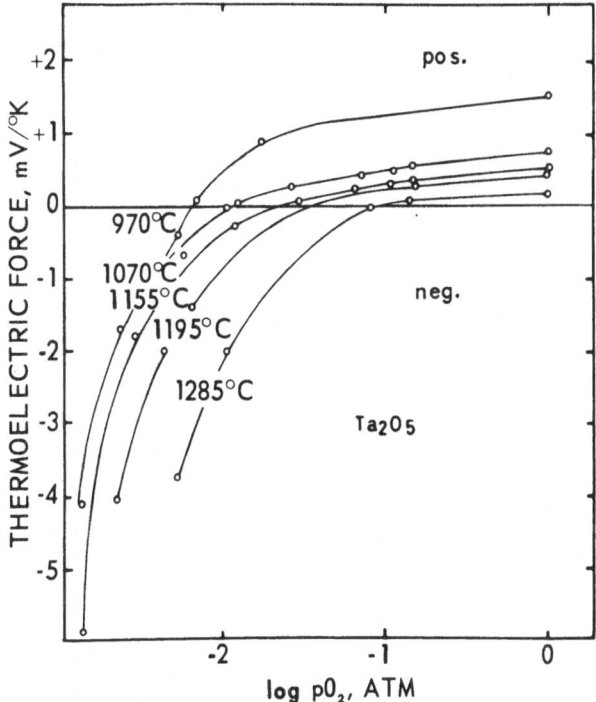

Figure 14. Thermoelectric force of Ta_2O_5 specimen as a function
of the logarithm of the partial pressure of oxygen [from Kofstad,
J. Electrochem. Soc. **109** (1962) 776].

counterelectrode. The direction of rectification, if present, is normally the same. This is consistent with a *p–n* junction, present
throughout in the oxide, being responsible for rectification.

Variation of capacitance with bias voltage analogous to the
effect with *p–n* junctions has been observed.[137,138]

2. Experimental Results on Electronic Current

(i) Low-Field Ohmic Region

Ohm's law (Figure 15) was observed by Mead[139] at room
temperatures with films of anodic tantalum pentoxide up to quite
high field strengths and over several decades of current (e.g., to
10^{-7} A/cm^2 at 0.5×10^6 V/cm with 110 Å films). Mead favored

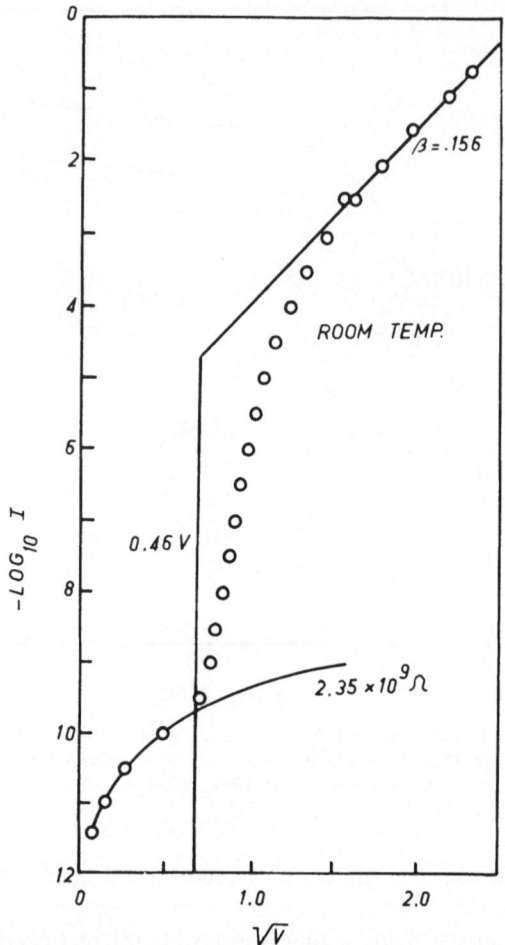

Figure 15. Typical voltage–ampere characteristic of Ta–Ta$_2$O$_5$–Au diode. Contact area about 10^{-3} cm^2. Note Ohmic region at low voltages and exp($V^{\frac{1}{2}}$) region at high voltages. Mead compared the transition voltage (0.46 V) with the difference in the work functions of the metals (110 Å oxide thickness) [from Mead, *Phys. Rev.* **128** (1962) 2088].

bulk control for this region even with very thin films (110 Å). This is in line with his view on interface *vs.* bulk control for the higher ranges of field. In this low field region, Mead suggested that the fact that Ohm's law was found up to voltages much greater than kT/e ruled out barrier control. It seems to us that the voltage range could exceed several kT/e without eliminating injection control because, if space charge is negligible, the field required at the barrier to produce a given current must be continued right through the film. The range of current over which Ohm's law is obeyed might seem to be more significant. One would expect for barrier control, with a barrier of fixed width, a relation of the general form: $I = I_0\{\exp(\alpha E/kT) - \exp[(1 - \alpha)E/kT]\}$. The linear region would then extend only over currents up to the order of I_0. The difficulty is to estimate I_0. It seems surprising that the bulk properties rather than the injection process should control the current down to films as thin as 110 Å. However, this could indeed happen with a trap density which is high enough to provide sufficient space charge for the field to change rapidly over a few Å. Mead's suggestion was that because of their structure, the amorphous oxides have a high density of localized states with energies in the forbidden gap. He suggested also that thermally excited hopping from state to state controls the current at room temperatures.

(ii) Schottky-Type Characteristics

Several authors[140-144] have found a region of field at room temperatures, with not too thin films, in which a Schottky or Poole–Frenkel law, $\log(I/I_0) = \beta E^{\frac{1}{2}}$, describes the current–voltage characteristic (Figures 15–17). Much of the discussion in the literature is concerned with the problem of distinguishing between these two mechanisms. The same law was found, and the same controversy developed, many years ago for "pre-breakdown currents" in macroscopic insulators.[37]

Below the region in which one of the above mechanisms is thought to apply, and above the Ohm's law region, a transition region has been reported in which the current increases more rapidly with field. (There will of course always be a steeper part for small E on a $\log I$ vs. $E^{\frac{1}{2}}$ plot, as Ohm's law takes over.) Mead's plots (Figures 15 and 17) show that the current in the transition region could be fitted with the Schottky or Poole–Frenkel law with a much bigger β than in the region at higher fields, to which Mead considers this law to apply. With films of 110 Å this transition region occupied five decades of current. With 870 Å, the transition

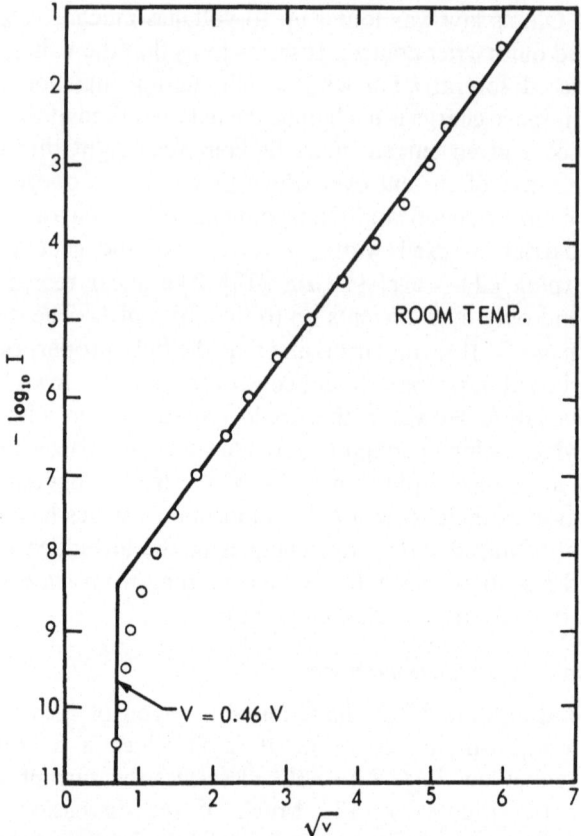

Figure 16. Voltage–ampere characteristic for thicker films show-
ing $\exp(V^{\frac{1}{2}})$ region and transition at 0.46 V. (870 Å oxide thickness,
area of contact about $10^{-3}\,\text{cm}^2$) [from Mead, *Phys. Rev.* **128**
(1962) 2088].

region occupied only two decades. We shall return to this point
below. Standley and Maissel[143] suggested that the rapid rise could
be due to a trap-filled limit effect. They confirmed that this was
possible by an order of magnitude calculation. As shown below,
the current in the Schottky region does not show space-charge
effects (i.e., I is a function of E and not of V^2/L^3). This does not rule
out their explanation. The reduction in the resistance of the oxide

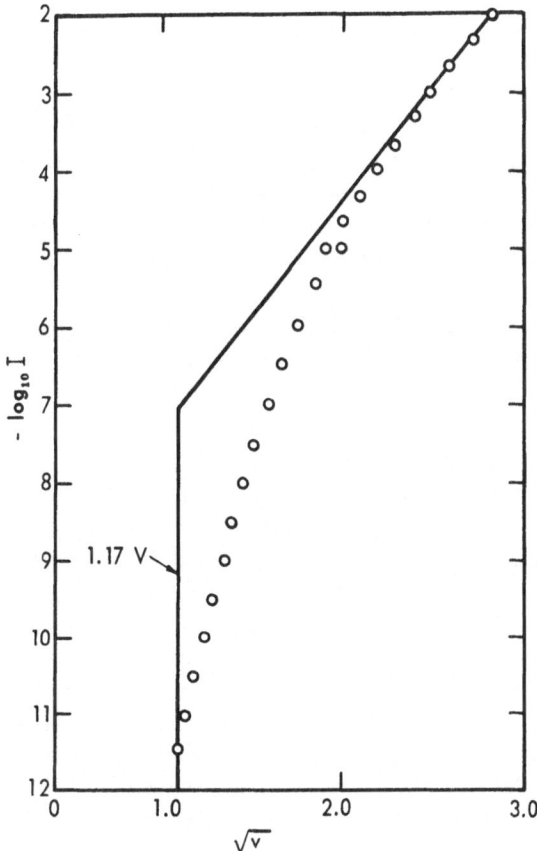

Figure 17. Voltage–ampere characteristic of Ta–Ta$_2$O$_5$–Pt diode showing transition at higher voltage. 1.17 V is the difference in work functions. Contact area about 10^{-3} cm^2 [from Mead, *Phys. Rev.* **128** (1962) 2088].

as the traps filled could leave the Schottky injection mechanism in control; Standley and Maissel favor the Schottky over the Poole–Frenkel mechanism. Alternatively, some form of avalanche effect rather than a trap-filled limit effect could be responsible for the rapid rise in the transition region. The avalanche effect could make the oxide highly conducting and the current would then be limited by injection. Mead considered that the current is low until the field due to the contact potential difference is cancelled by the applied field. A discussion of this appears below.

One would like to be sure that the current is due to uniform conduction, and not to defects or weak places in the films. Some of the earlier data, in our opinion, undoubtedly refer to nonuniform conduction. The Schottky law is still observed in some such cases. It is difficult to compare results from different laboratories because of the variation in conditions under which the oxide was made and the counterelectrodes applied, and because of uncertainties about contact areas, and other details. For example, the oxide will be heated by radiation from the evaporation source, and it is known that a rise of a few 10°C has an appreciable effect on Ta_2O_5 films.

Mead (loc cit. p. 2089) takes an observed variation between samples in the temperature dependence as evidence against the Schottky mechanism and seems to imply that variation in trap density would cause variability between specimens to be an expected feature for bulk control. We suggest that such localized states might be expected to correlate with the concentration of mobile ions in ionic conduction experiments. It would be interesting to see if any of the variation between specimens could be correlated with differences in current density of formation or to differences in later annealing. The ionic conductivity is very reproducible and, therefore, the oxide material, as first made, is probably reproducible. However, the variations suggested by Mead could arise, as indicated above, from variations in treatment, particularly annealing, subsequent to formation.

Standley and Maissel have developed a current-reversal technique which is applied during the formation of the films and which reduces the spread of behavior, apparently due to the elimination of some of the worst weak places in the film. However, a high density of defects was indicated by Vermilyea's work (section 2.) and a high density would give good reproducibility. Because one knows that defects are present and that they make the electronic current irreproducible under some circumstances (in contrast with the ionic current which is very reproducible) caution is required, and this applies to all regions of the I-V-T-d characteristics. The obvious test is to vary the area of the contact. The probability of avoiding a defect, if these are present in concentration one per area A_0, is $\exp(-A/A_0)$. Very small contact areas indeed are

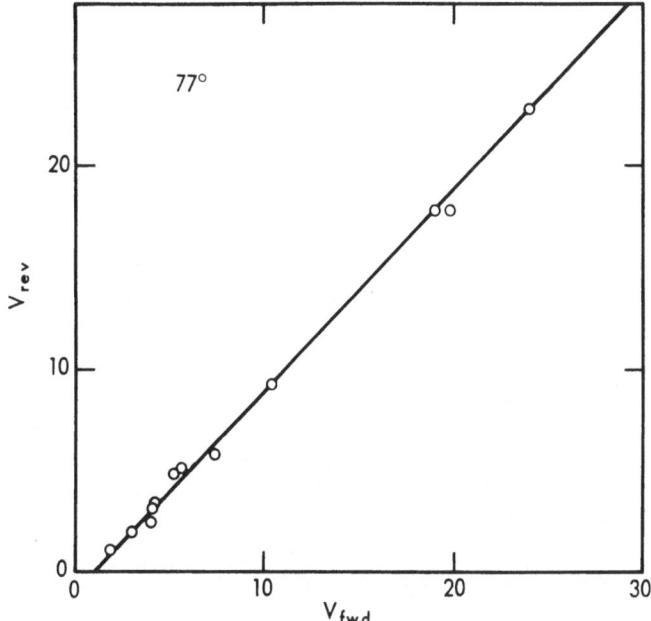

Figure 18. Comparison of voltage required for 1 mA in forward and reverse directions for a number of Ta–Ta$_2$O$_5$–Au diodes. The straight line shown is predicted by a bulk mechanism. Area of contact about 10^{-3}cm^2 [from Mead, *Phys. Rev.* **128** (1962) 2088].

required. In tests of this kind we have observed appreciable reductions in current density at a given voltage as the area is decreased.

It is interesting to compare currents in the wet and dry systems. For an example, one of Mead's Figures (our Figure 16) shows 0.1 A at room temperature for 10^{-3} cm^2 at 36 V and 870 Å, i.e., 10^2 A/cm^2 at 5×10^6 V/cm, with tantalum negative. According to Figure 18 the asymmetry was small. In the wet system with tantalum positive one would observe about 10^{-6} A/cm^2 at this field and this current would be nearly all ionic.

The *I vs. E* relation should be independent of thickness for both bulk and injection control except for the expected region of thickness where the transition from injection to bulk control occurs. Mead (Figure 19) found a fairly but not exactly complete lack of dependence on thickness of β in $\log(I/I_0) = \beta E^{\frac{1}{2}}$ for the

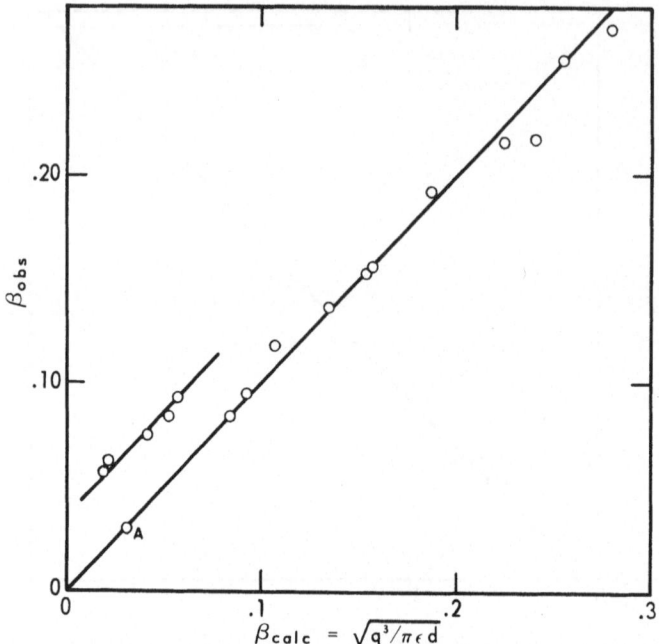

Figure 19. Comparison of observed and calculated $\log I$ vs. $V^{\frac{1}{2}}$ slopes. Points below $\beta_{calc} = 0.08$ correspond to the upper branch of the line. β in this figure corresponds to $kT\beta/d^{\frac{1}{2}}$ in the text [from Mead, *Phys. Rev.* **128** (1962) 2088].

thicker films. (He observed an anomalous jump in plots of V for given I vs. $1/C$ which we shall not discuss.) This rules out space-charge limited currents which might simulate a Schottky law near the trap filled limit.

The value of β was taken by Mead as differentiating between the Schottky and Poole–Frenkel mechanisms. Mead used the audio-frequency dielectric constant which is reproducible but which depends on the conditions under which the anodic oxide is made. He obtained a good fit to the Poole–Frenkel slope. One might expect that the dielectric constant effective in the image or coulombic force equation should be some form of high-frequency value.[102] This might be very considerably lower than the audio-frequency value [$\varepsilon(1 \text{ Kc})s = 28$, $n(NaD) = 2.2$]. This would give a much greater β. It might even give a value which would fit the slope in the wide transition region which is present in Mead's plots for thinner films and which was mentioned above. For a velocity of 10^6 cm/sec which corresponds to a kinetic energy of about kT, a barrier of width 100 Å would be traversed in 10^{-12} sec. Therefore, it is doubtful whether the dielectric constant should be the value above or the value below the large change at reststrahlen frequencies. With SiO films we find reasonable agreement if the optical value is used. Also, the image force and coulombic force assumptions are not to be expected to be very close approximations. One might argue that the factor of 2 difference in slope between the Poole–Frenkel and Schottky laws is not really very significant.

The temperature dependence has also been used to discriminate between the Poole–Frenkel and Schottky mechanisms. The T^2 term present in the usual form of the Schottky law and absent in the Poole–Frenkel law seems really to be of little significance, either experimentally or theoretically, as a means of deciding between these mechanisms. The pre-exponential factor of the Schottky theory contains the Richardson constant but this constant can hardly be expected to have its usual theoretical value. In any case it should contain a factor for the probability of transmission through the barrier for an electron incident on the barrier region with energy greater than the barrier height. This factor will depend on the precise configuration of the barrier region, which determines whether the electron will lose its energy before getting over the barrier. Furthermore, the potential-energy barrier approach is only an artifice to obtain a model amenable to calculation.

The value of β is predicted by both theories to be inversely proportional to temperature. Mead reported excellent agreement for 300 and 195°C as shown by Figure 19, which includes points from experiments at both temperatures. Mead does not refer to any allowance for the temperature dependence of ε.

The effect of the particular metal used for the counter-electrode in the Schottky-law region has been investigated by several authors. Mead, as stated above, has suggested that the rapid rise to the Poole–Frenkel region occurs when the electrostatic field in the oxide due to the work function difference is just cancelled by the applied field. The voltage at which the rapid rise occurs was thus equated to the difference in work functions of the clean metal surfaces. It would seem that this picture should apply only to the thin-film case in which space charge is negligible, since in a thick enough film the field due to the contact PD is confined to the space-charge regions near the interfaces, the potential being constant in the middle part of the film with no voltage applied. If the injection processes are considered to have negligible control, the applied Fermi-level difference will appear wholly as a field in this middle part of the film. Since Mead assumed the Poole–Frenkel mechanism, which is, of course, a bulk-control process, his suggestion about work functions seems to require more justification than was originally given. Standley and Maissel tested the hypothesis with eight metals for counterelectrode, including the platinum, gold, and aluminum tested by Mead. They found no correlation between the difference in work function between tantalum and the counter-electrode metal and the voltage at which appreciable current first appeared (10^{-10} A for 0.0013 cm^2 contact area). Standley and Maissel found that counterelectrode metals with lower work functions gave currents in the range 0.1 mA/cm^2 for lower voltages. This was with the counterelectrode negative and, surprisingly, to some extent with the counterelectrode positive (Figures 20 and 21). The behavior with the counterelectrode negative indicated two groups of metals for which the work function was greater or smaller,

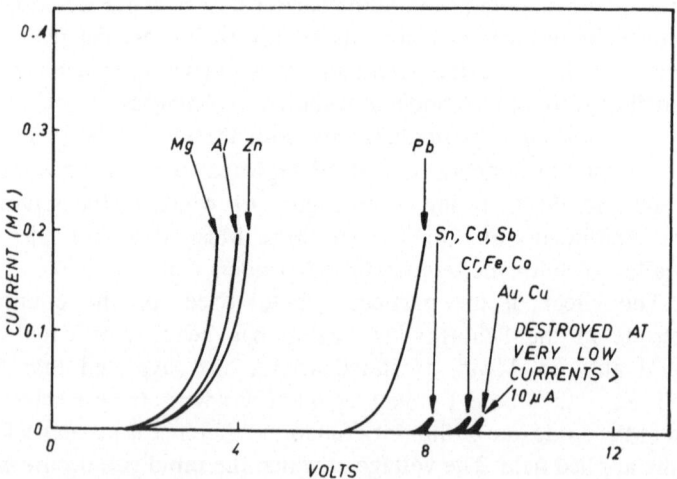

Figure 20. Current–voltage curves for various counterelectrode metals [from Standley and Maissel, *J. Appl. Phys.* **35** (1964) 1530].

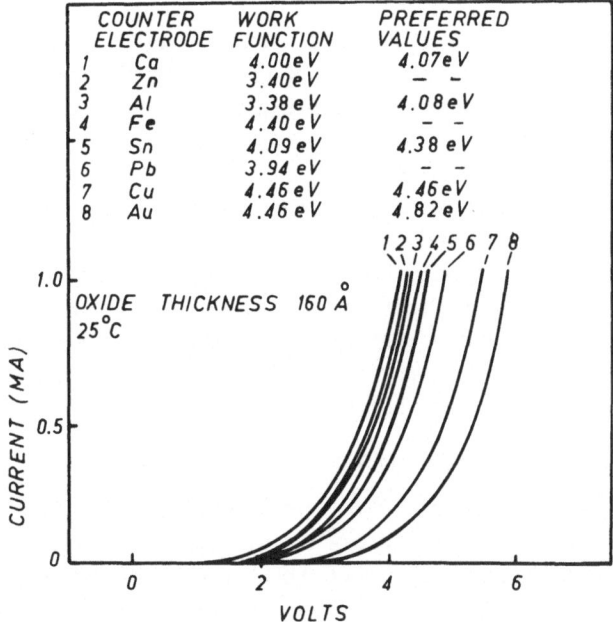

COUNTER ELECTRODE		WORK FUNCTION	PREFERRED VALUES
1	Ca	4.00 eV	4.07 eV
2	Zn	3.40 eV	— —
3	Al	3.38 eV	4.08 eV
4	Fe	4.40 eV	— —
5	Sn	4.09 eV	4.38 eV
6	Pb	3.94 eV	— —
7	Cu	4.46 eV	4.46 eV
8	Au	4.46 eV	4.82 eV

Figure 21. Current–voltage curves for various counterelectrode metals. Contact area 1.3×10^{-3} cm^2. (Counterelectrode positive with respect to tantalum) [from Standley and Maissel, *J. Appl. Phys.* **35** (1964) 1530].

respectively, than 4 eV. The group with $\phi < 4$ eV gave nearly symmetrical *I–V* characteristics. The dependence on counterelectrode with the counterelectrode negative would tend to indicate the Schottky rather than the Poole–Frenkel mechanism.

Simmons[130] treated tunneling and Schottky emission through very thin films with a double-image force barrier. With thin enough films the current depended on the work function of the positive electrode, as in Standley and Maissel's work.

(iii) Exponential Region

Advani, Gottling, and Osman,[145] and also Hacskaylo[142] have observed a region of field where $\log(I/I_0) = \beta E$. This law would correspond to thermal promotion over a barrier whose activation distance (to borrow a useful term from the ionic-conduction terminology) is not dependent on field. Hacskaylo claims that the voltage at which the transition occurs, corresponds to a condition where the vacuum level near the positive metal is level with the bottom of the conduction band of the insulator. Jacobs[146] considered an avalanche mechanism to explain the same law with MgO.

(iv) Fowler–Nordheim-Type Law

At high fields and low temperatures, where the thermal activation Schottky and Poole–Frenkel mechanisms are frozen out, and for films too thick for tunneling from metal to metal, a Fowler–Nordheim law is expected. Mead has observed this relationship for Ta_2O_5 films, it could be due either to tunneling from the metal into the conduction band or from localized states to the conduction band. Mead favored the latter case.

(v) Tunneling from Metal to Metal

With films less than 100 Å, tunneling from metal to metal tends to control the current. Most of the work is for air-formed oxide. It has been confirmed by Fisher and Giaever[147] that the resistance increases exponentially with increasing thickness (Figure 22) and is relatively insensitive to temperature (Figure 23), as is required by the tunneling mechanism. Variation of contact area gave currents roughly proportional to the area, which was taken to indicate that defects were not involved. At low fields the $I–V$ relation is Ohmic; as the field is increased, the current increases roughly exponentially with field (Figure 24). The Fowler–Nordheim law region was not reached because of breakdown.

Fisher and Giaever found that their results could be accounted for quantitatively in terms of Holm's theory by taking an effective mass of $m/9$, which is a surprising effective mass for such wide-band materials.

A particularly interesting effect was found by Giaever and Megerle[148] to occur with superconducting metal counterelectrodes. The energy gaps give negative resistance regions analogous to that for the $p–n$ junction Esaki tunnel diode.

With such extremely thin films, it is difficult to be sure that one has a known thickness of uniform insulator. The usual method has been to determine the thickness from capacitance measurements, assuming a value of the dielectric constant. The true dielectric constant may not be the same as that for films with more than a

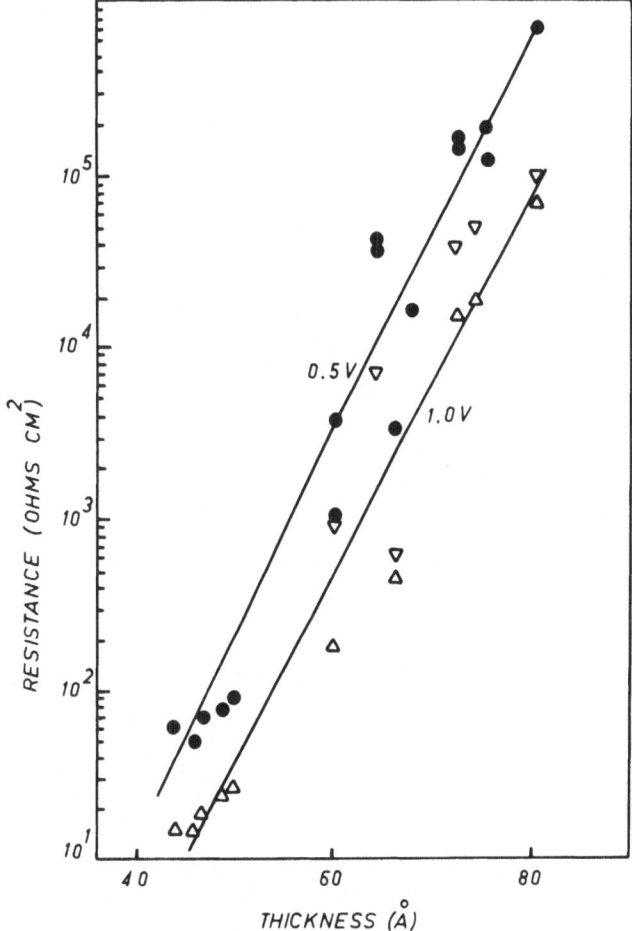

Figure 22. Exponential increase of resistance with thickness. (Al_2O_3
films formed in air at room temperature and then at 400°C) [from
Fisher and Giaever, *J. Appl. Phys.* **32** (1961) 172].

few atomic layers of dielectric. Mead[149] observed that plots of the
voltage for a given current *vs.* reciprocal capacitance did not
extrapolate to zero reciprocal capacitance. He concluded that the
field penetrated into the metal an appreciable distance (Å). The
possible formation of *p*- and *n*-type layers in the oxide has been

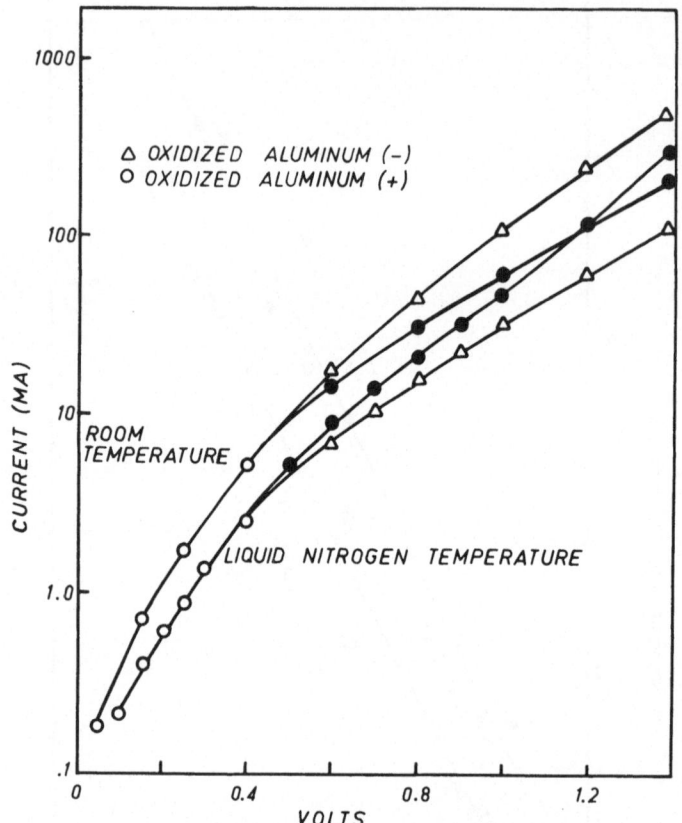

Figure 23. Rectification, observed with films formed at elevated temperatures, reverses sign with temperature [from Fisher and Giaever, *J. Appl. Phys.* **32** (1961) 172].

discussed by Fisher and Giaever[147], Pollack and Morris,[150] and by others. Such concepts seem unrealistic in films only 50 Å or less thick. In the interpretation of the data using the theoretical formula, the dielectric constant should again presumably be some form of high-frequency value since the transit time of the electron is now very short. Possibly a value in the optical range should be used for the image force. Handy[151] has found a correlation between tunneling resistance and the "atomic" radius of the counterelectrode

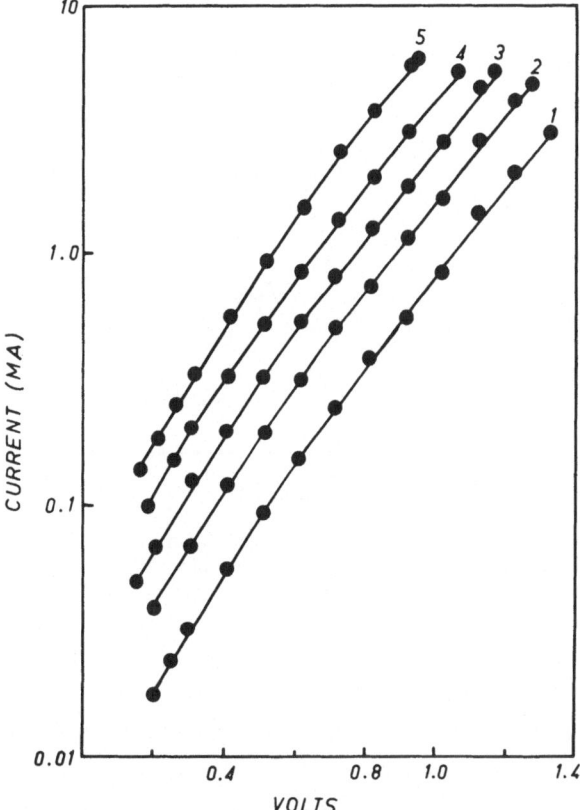

Figure 24. At higher voltages, the current through thin oxide films (~ 47 Å) increases exponentially with voltage. Curves are shown for five films with areas in the proportions $5:4:3:2:1$ (in units of 4.7×10^{-3} cm^2) as indicated [from Fisher and Giaever, *J. Appl. Phys.* **32** (1961) 172].

metal. Harris[144,152] investigated currents in BeO films thicker than used by Simmons, Unterkofler and Allen[153] and Chow[154] who observed tunneling. He suggested that the small temperature dependence which he observed did not indicate tunneling but was consistent with Schottky emission where the current is carried by a multitude of weak places in the film and heating occurs in such a way that the temperature is independent of the surroundings. Cohen[155] has taken issue with this view.

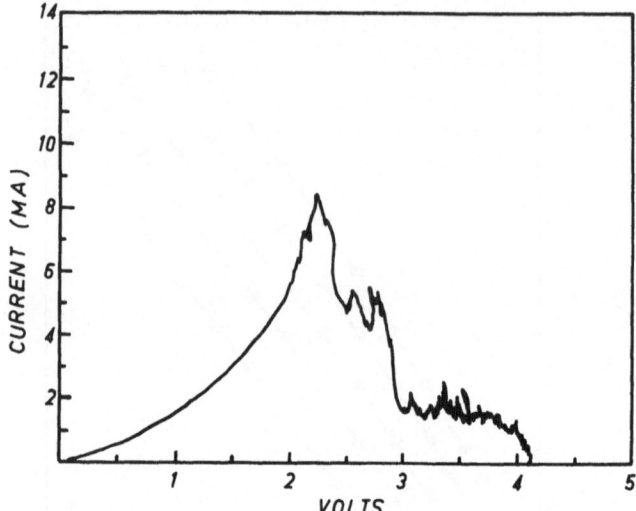

Figure 25. Tracing of $X-Y$ recorder plot of the establishment of conductivity in a 350 Å aluminum oxide film. Au = +, Al = −. Contact area 1–2 mm^2 [from Hickmott, *J. Appl. Phys.* **33** (1962) 2669].

(vi) Negative Resistance

Hickmott[156–158] found that metal/anodic oxide/metal sandwiches exhibit a voltage controlled negative resistance if they are first subjected to a "forming" process. The development of conductivity in a 350 Å anodic Al_2O_3 film with the gold counterelectrode positive is shown in Figure 25. Negative resistance with the gold negative is illustrated in Figure 26. The negative resistance can be observed with either the aluminum or the gold negative, with no qualitative difference in characteristics. The polarity of the forming voltage has again only secondary effects. The forming process need not be performed in vacuum, though early attempts to form in air were unsuccessful, but the best characteristics are obtained by formation in vacuum, and various changes in the characteristics are observed when gases are introduced subsequent to the formation process.

The reduction in the conductivity in going from peak to valley is fast ($<10^{-6}$ sec) but the re-establishment of the conductivity is slow (seconds). Negative resistance can be observed with triode

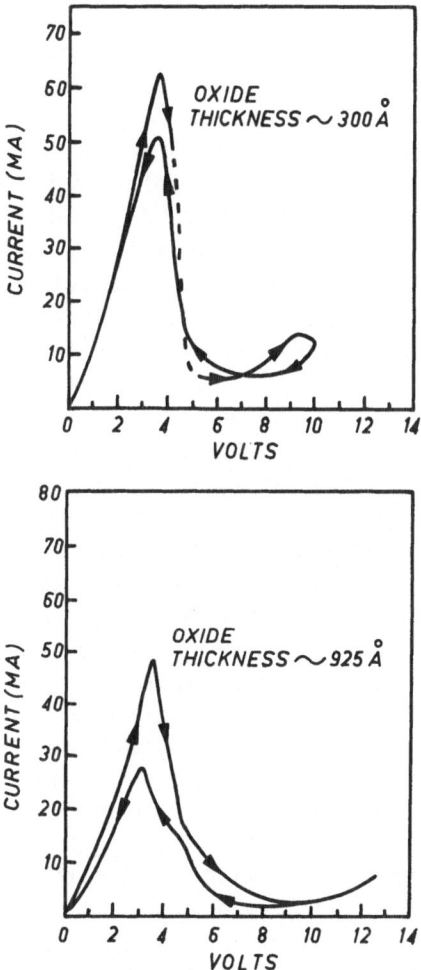

Figure 26. Dependence of direct current–
voltage characteristics of Al–Al$_2$O$_3$–Au
sandwich on aluminum oxide thickness.
Al = +, Au = −. Contact area 1–2 mm^2
[from Hickmott, *J. Appl. Phys.* **33** (1962)
2669].

structures: metal/oxide/metal/oxide/metal. These devices give
information on the distribution of the field. The experiments
showed that a large field is not required, and is not necessarily

present, near the negative electrode. Thus field emission from this electrode does not necessarily control the characteristic. Negative resistance has been investigated by several authors,[156–163,141] and has been observed with Al_2O_3, SiO, Ta_2O_5, ZrO_2, TiO_2, MgO + Al_2O_3, and MgO. A correlation of the characteristics can be made with the dielectric constant (audio) and with the band gap of the oxide. Forming aluminum in $KHSO_4$ + NH_4HSO_4 eutectic gives films containing a large concentration of some sulfur-bearing entity, and this doping gives interesting differences in behavior from films formed in boric acid + borate aqueous solutions and other electrolytes.

The current below the peak shows the $I \propto V^2$ dependence which is characteristic of space charge limited currents (Figure 27). The current is a function of (voltage)^2, not of (voltage/total thickness of oxide). It is nearly independent of temperature from room temperature to 3°K. Electron emission into vacuum and electroluminescence have been observed and indicate energetic electrons.

The details of the experimental behavior are so complicated that the reader must be referred to the originals for further details.

Hickmott has discussed the possible mechanisms in considerable detail. The tentative theory which he has advanced is, briefly, that the production of conductivity in the forming process is due to the development of impurity-band conduction,[101] in a concentrated band of impurity states which are ionized in the forming process. The current is then space-charge limited. The impurity-band nature of the conduction is supposed to account for the small temperature dependence. To account for the negative resistance it is supposed that as the field is increased the impurity-band states are filled by electrons tunneling from valence band or lower impurity states. The process involved in the development of the impurity band conduction is somewhat uncertain.

Ridley[168] has, of course, predicted that negative resistance should be a general property of photoconductors showing field-enhanced trappings.

Current-controlled negative resistance capable of sustaining oscillations at 10 Mcps has been observed by Geppert,[164] Chopra,[165] and Beam and Armstrong,[166] in oxide films on niobium. A forming process involving the application of pulses of various lengths is needed to develop the negative resistance. Beam and

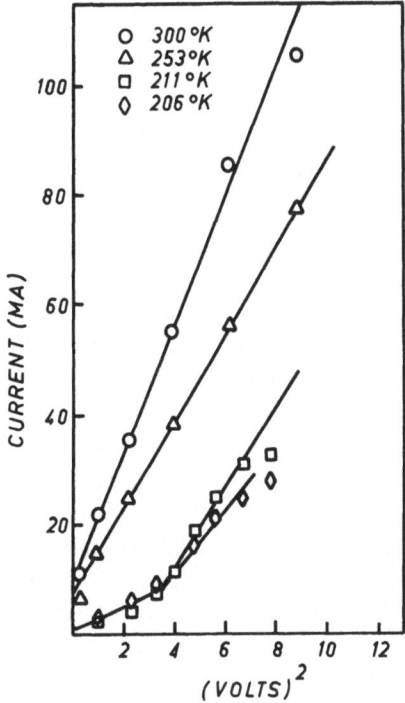

Figure 27. Space-charge-limited currents in a 450 Å aluminum oxide film at different temperatures. Contact area 1–2 mm² [from Hickmott, *J. Appl. Phys.* **33** (1962) 2669].

Armstrong suggested that all but a layer 25 Å or so in thickness of oxide near the niobium is made conducting by some effect of the forming process. The current is then controlled by field emission of electrons from the negative niobium into this thin insulating layer. The field across this layer is enhanced by an amount proportional to a space charge produced by the impact ionization of niobium atoms. It was assumed that the amount of space charge is proportional to the current flowing. This gives a negative resistance which terminates when all the niobium atoms in the effective region are ionized. The *I–V* characteristics were fitted by subtracting a field proportional to the current from the Schottky-law field.

(vii) Dielectric Breakdown

By dielectric breakdown one usually means in practice a permanent modification of the insulator due to the passage of high current. Permanent damage will usually occur only if ionic conduction occurs or if the temperature rises. The damage is indicated either by a conductive path, as in the "forming" processes used to develop the negative resistance characteristics, or, more violently, by melting or evaporation of the counterelectrode and even of the oxide and some of the metal substrate as well. Dielectric breakdown is usually discussed[119] in terms of processes which give electronic currents which rise rapidly at critical fields, but such currents will result in permanent damage only if the heat which they generate is not removed fast enough, and the present films are thin enough to be cooled more effectively than bulk material. Breakdown, in practice, at least of the more violent kind, takes place locally. This is to be expected from the regenerative feedback due to the heat produced where the current is locally high, but in many cases it is undoubtedly due to defects in the film, so that true breakdown strengths are largely unknown. Electrolytic capacitors of the conventional types are operated at fields of up to several 10^6 V/cm but the oxide tends to be self-repairing if electrolyte is present. Thus the problem of breakdown becomes more acute in the devices which employ metal/oxide/metal and similar systems.

ACKNOWLEDGMENTS

We thank the Defence Research Board of Canada and the Sprague Electric Co. for financial support for our work in this field.

REFERENCES

[1] Young, *Anodic Oxide Films*, Academic Press Inc., New York, 1961.
[2] Vermilyea, *Advances in Electrochemistry*, Vol. 3, p. 211, Ed. P. Delahay, Interscience Publishers, Inc., New York, 1963.
[3] Calvert and Draper, *Can. J. Chem.* **40** (1962) 1943.
[4] Vermilyea, *J. Electrochem. Soc.* **104** (1957) 427.
[5] Vermilyea, *J. Electrochem. Soc.* **102** (1955) 207.
[6] Vermilyea, *J. Electrochem. Soc.* **104** (1957) 542.
[7] Guntherschulze and Betz, *Elektrolytkondensatoren*, Krayn, Berlin, 1937; 2nd ed. Cram, Berlin, 1952.
[8] Young, *Trans. Faraday Soc.* **55** (1959) 842.
[9] Vermilyea, *J. Electrochem. Soc.* **110** (1963) 250.

[10] Vermilyea, *J. Electrochem. Soc.* **110** (1963) 345.
[11] Vermilyea, *Acta Met.* **2** (1954) 476.
[12] Young, *Acta Met.* **5** (1957) 711.
[13] Vermilyea, *Acta Met.* **5** (1957) 113.
[14] Vermilyea, *J. Electrochem. Soc.* **104** (1957) 485.
[15] Masing, Orme, and Young, *J. Electrochem. Soc.* **108** (1961) 428.
[16] Young, *Proc. Roy. Soc. London, Ser. A* **258** (1960) 496.
[17] Young, *Acta Met.* **4** (1956) 101.
[18] Young, *J. Electrochem. Soc.* **111** (1964) 1289.
[19] Dewald, unpublished.
[20] Vermilyea, *J. Electrochem. Soc.* **104** (1957) 140.
[21] Kumagai and Young, *J. Electrochem. Soc.* **111** (1964) 1411.
[22] Young, *Proc. Roy. Soc. London, Ser. A* **244** (1958) 41.
[23] Heavens and Kelly, *Proc. Phys. Soc.* **72** (1958) 906.
[24] Claussen, personal communication.
[25] Frenkel, *Kinetic Theory of Liquids*, p. 40, Dover Publishing Inc., New York, 1955.
[26] Barrer, *Diffusion in and through Solids*, p. 300, Cambridge University Press, 1951.
[27] Young, *Proc. Roy. Soc. London, Ser. A* **263** (1961) 395.
[28] Young, *J. Electrochem. Soc.* **110** (1963) 589.
[29] Jost, *Diffusion in Solids, Liquids and Gases*, Academic Press Inc., New York, 1952.
[30] Paul and Warschauer, *Solids under Pressure*, pp. 43, 63, McGraw-Hill Book Co., New York, 1963.
[31] Young, *Trans. Faraday Soc.* **50** (1954) 153.
[32] Nelson and Thompson, *Proc. Roy. Soc. London, Ser. A* **259** (1961) 458.
[33] Vermilyea, *Acta Met.* **1** (1953) 282.
[34] Vermilyea, *Acta Met.* **3** (1955) 106.
[35] Verwey, *Physica* **2** (1935) 1059.
[36] Mott, *Trans. Faraday Soc.* **43** (1947) 429.
[37] Mott and Gurney, *Electronic Processes in Ionic Crystals*, Oxford University Press, 1959.
[38] Cabrera and Mott, *Rept. Prog. Phys.* **12** (1948–9) 163.
[39] Bray, Jacobs, and Young, *Proc. Phys. Soc.* **71** (1958) 405.
[40] Bean, Fisher, and Vermilyea, *Phys. Rev.* **101** (1956) 551.
[41] Dewald, *J. Phys. Chem. Solids* **2** (1957) 55.
[42] Young, *Trans. Faraday Soc.* **52** (1956) 502, 515.
[43] Vermilyea, *J. Electrochem. Soc.* **102** (1955) 655.
[44] Dreiner, *J. Electrochem. Soc.* **111** (1964) 1350.
[45] Benjamini, Duffek, Mylroie, and Schulenburg, *J. Electrochem. Soc.* **110** (1963) 266C.
[46] Condit and Holt, *J. Electrochem. Soc.* **111** (1964) 1192.
[47] Amsel and Samuel, *J. Phys. Chem. Solids* **23** (1962) 1707.
[48] Graham, Davies, and Brown, *Bull. Am. Phys. Soc. Ser. II* **7** (1962) 491.
[49] Graham, Brown, Davies, and Pringle, *Can. J. Phys.* **41** (1963) 1686.
[50] Davies, Pringle, Graham, and Brown, *J. Electrochem. Soc.* **109** (1962) 999.
[51] Davies, Friesen, and McIntyre, *Can. J. Chem.* **38** (1960) 1526.
[52] Renshaw, *J. Electrochem. Soc.* **108** (1961) 185.
[53] Pringle and Davies, Extended Abstracts, Theoretical Division, Vol. 1, No. 1, 63, Electrochemical Society Meeting, Pittsburgh, April, 1963.
[54] Davies and Domeij, *J. Electrochem. Soc.* **110** (1963) 849.

[55] Davies, Domeij, Pringle, and Brown, to be published.

[56] Domeij, Brown, Davies, and McCargo, *Can. J. Phys.* **42** (1964) 1624.

[57] Cheseldine, *J. Electrochem. Soc.* **111** (1964) 1128.

[58] Ligenza and Spitzer, *J. Phys. Chem. Solids* **14** (1960) 131.

[59] Pliskin and Gnall, *J. Electrochem. Soc.* **111** (1964) 872.

[60] Young, *Can. J. Chem.* **37** (1959) 276.

[61] Smyth, Shirn, and Tripp, *J. Electrochem. Soc.* **111** (1964) 1331.

[62] Smyth, Shirn, and Tripp, *J. Electrochem. Soc.* **110** (1963) 1264.

[63] Pavlovic, *J. Chem. Phys.* **40** (1964) 951.

[64] Kofstad, *J. Electrochem. Soc.* **109** (1962) 776.

[65] Young, *Trans. Faraday Soc.* **53** (1957) 841.

[66] Dignam, *Can. J. Chem.* **42** (1964) 1155.

[67] Dignam, Extended Abstracts, Theoretical Division, Vol. I., No. 1, 90, Electrochemical Society Meeting, Pittsburgh, April 1963.

[68] Vermilyea, General Electric Research Labs. Report No. 64-RL-3619M, March 1964.

[69] Vermilyea, General Electric Research Labs. Report No. 64-RL-3682M, June 1964.

[70] McCrackin, Passaglia, Stromberg, and Steinberg, *J. Res. Nat. Bur. Std. A.* **67** (1963) 363.

[71] Archer, *J. Opt. Soc. Am.* **52** (1962) 970.

[72] Zaininger and Revesz, *R.C.A. Rev.* **25** (1964) 85.

[73] Archer, *J. Electrochem. Soc.* **104** (1957) 619.

[74] Claussen and Flower, *J. Electrochem. Soc.* **110** (1963) 983.

[75] Menard, *J. Opt. Soc. Am.* **52** (1962) 427.

[76] Pimbley and Macqueen, *J. Phys. Chem.* **68** (1964) 1101.

[77] Claussen, *J. Electrochem. Soc.* **111** (1964) 646.

[78] Bockris, Devanathan, and Reddy, *Proc. Roy. Soc. London, Ser. A* **279** (1964) 327.

[79] Schmidt and Michel, *J. Electrochem. Soc.* **104** (1957) 230.

[80] Politycki and Fuchs, *Z. Naturforsch.* **14A** (1959) 271.

[81] Duffek, Mylroie, and Benjamini, *J. Electrochem. Soc.* **111** (1964) 1042.

[82] Duffek, Benjamini, and Mylroie, *J. Electrochem. Soc.* **111** (1964) 62C.

[83] Rabinovitch and Borrer, Extended Abstracts, Electronics Division, Vol. II, No. 1, 220, Electrochemical Society Meeting, Los Angeles, May 1962.

[84] Lewis, *J. Electrochem. Soc.* **111** (1964) 1007.

[85] Van Tassel and Haberecht, Extended Abstracts of Joint Symposium, 93. Electrochemical Society Meeting, Pittsburgh, April 1963.

[86] Haas, *J. Electrochem. Soc.* **109** (1962) 1192.

[87] Schmidt and Owen, *J. Electrochem. Soc.* **111** (1964) 682.

[88] Waring and Benjamini, *J. Electrochem. Soc.* **111** (1964) 1256.

[89] Dubrovskii, Mel'nik, and Odynets, *Russian J. Phys. Chem.* **36** (1962) 1183.

[90] Zwerdling and Sheff, *J. Electrochem. Soc.* **107** (1960) 338.

[91] Wales, *J. Electrochem. Soc.* **110** (1963) 914.

[92] Vasicek, *Optics of Thin Films*, North-Holland Publishing Co., Amsterdam, 1960.

[93] Wales, *J. Electrochem. Soc.* **111** (1964) 478.

[94] Turner, The Electrochemistry of Semiconductors, Ed. P. J. Holmes, p. 178, Academic Press Inc., London, 1962.

[95] Dewald, *J. Electrochem. Soc.* **104** (1957) 244.

[96] Venables and Broudy, *J. Appl. Phys.* **30** (1959) 1110.

[97] Harkness and Young, unpublished work.

[98] Mueller and Jacobson, *J. Appl. Phys.* **35** (1964) 1524.

[99] Frenkel, *Tech. Phys. USSR* **5** (1938) 685.

[100] Ioffe and Regel, *Progr. Semicond.* **4** (1960) 239.

[101] Mott and Twose, *Advan. Phys.* **10** (1961) 107.

[102] Henisch, *Rectifying Semiconductor Contacts*, Oxford University Press, 1955.

[103] Mott and Gurney, *Electronic Processes in Ionic Crystals*, Oxford University Press, 1959.

[104] Simmons, *Phys. Rev. Letters* **10** (1963) 10.

[105] Wright, *Nature* **182** (1958) 1296.

[106] Smith and Rose, *Phys. Rev.* **97** (1955) 1531.

[107] Rose, *Phys. Rev.* **97** (1955) 1538.

[108] Skinner, *J. Appl. Phys.* **26** (1955) 498.

[109] Lampert, *Phys. Rev.* **103** (1956) 1648.

[110] Suits, *J. Appl. Phys.* **28** (1957) 454.

[111] Ruppel, *Helv. Phys. Acta* **31** (1958) 311.

[112] Wright, *Solid-State Electron.* **2** (1961) 165.

[113] Zuleeg and Muller, *Solid-State Electron.* **7** (1964) 575.

[114] Lampert and Edelman, *J. Appl. Phys.* **35** (1964) 2971.

[115] Rose, *J. Appl. Phys.* **35** (1964) 2664.

[116] Schottky, *Z. Physik* **15** (1914) 872.

[117] Chynoweth, *Prog. Semicond.* **4** (1960) 95.

[118] Frenkel, *J. Exptl. Theoret. Phys. USSR* **8** (1938) 1893.

[119] O'Dwyer, *The Theory of Dielectric Breakdown of Solids*, Oxford University Press, 1964.

[120] Fowler and Guggenheim, *Statistical Thermodynamics*, p. 476, Cambridge University Press, 1939.

[121] Holm, *J. Appl. Phys.* **22** (1951) 569.

[122] Frenkel, *Phys. Rev.* **36** (1930) 1604.

[123] Sommerfeld and Bethe, *Handbuch der Physik*, Vol. 24–2, p. 432, Springer Verlag, Berlin, 1933.

[124] Holm and Kirschstein, *Z. Tech. Phys.* **16** (1935) 488.

[125] Harrison, *Phys. Rev.* **123** (1961) 85.

[126] Bardeen, *Phys. Rev. Letters* **6** (1961) 57.

[127] Stratton, *J. Phys. Chem. Solids* **23** (1962) 1177.

[128] Simmons, *J. Appl. Phys.* **34** (1963) 1793.

[129] Simmons, *J. Appl. Phys.* **34** (1963) 2581.

[130] Simmons, *J. Appl. Phys.* **35** (1964) 2472.

[131] Hartman, *J. Appl. Phys.* **33** (1962) 3427.

[132] Hartman, *J. Appl. Phys.* **35** (1964) 3283.

[133] Tantraporn, *Solid-State Electron.* **7** (1964) 81.

[134] Geppert, *J. Appl. Phys.* **33** (1962) 2993.

[135] Sasaki, *J. Phys. Chem. Solids* **13** (1960) 177.

[136] Ishikawa, Sasaki, Seki, and Inowaki, *J. Appl. Phys.* **34** (1963) 867.

[137] Huber and Rottersman, *J. Appl. Phys.* **33** (1962) 3385.

[138] Magill, *Solid-State Electron.* **6** (1963) 531.

[139] Mead, *Phys. Rev.* **128** (1962) 2088.

[140] Emtage and Tantraporn, *Phys. Rev. Letters* **8** (1962) 267.

[141] Pollack, *J. Appl. Phys.* **34** (1963) 877.

[142] Hacskaylo, *J. Appl. Phys.* **35** (1964) 2943.

[143] Standley and Maissel, *J. Appl. Phys.* **35** (1964) 1530.

[144] Harris, *J. Appl. Phys.* **35** (1964) 268.

[145] Advani, Gottling, and Osman, *Proc. I.R.E.* **50** (1962) 1530.

[146] Jacobs, *Phys. Rev.* **91** (1953) 804.

[147] Fisher and Giaever, *J. Appl. Phys.* **32** (1961) 172.
[148] Giaever and Megerle, *Phys. Rev.* **122** (1961) 1101.
[149] Mead, *Phys. Rev. Letters* **6** (1961) 545.
[150] Pollack and Morris, *J. Appl. Phys.* **35** (1964) 1503.
[151] Handy, *Phys. Rev.* **126** (1962) 1968.
[152] Harris, *J. Appl. Phys.* **35** (1964) 3057.
[153] Simmons, Unterkofler and Allen, *Appl. Phys. Letters* **2** (1963) 78.
[154] Chow, *J. Appl. Phys.* **34** (1963) 2918.
[155] Cohen, *J. Appl. Phys.* **35** (1964) 3056.
[156] Hickmott, *J. Appl. Phys.* **33** (1962) 2669.
[157] Hickmott, *J. Appl. Phys.* **35** (1964) 2118.
[158] Hickmott, *J. Appl. Phys.* **35** (1964) 2679.
[159] Kreynina, Selivanov, and Shumskaia, *Radio Eng. Elec. Phys.* **5** (1960) 8, 219.
[160] Kreynina, *Radio Eng. Elec. Phys.* **7** (1962) 166.
[161] Kreynina, *Radio Eng. Elec. Phys.* **7** (1962) 1949.
[162] Kanter and Feibelman, *J. Appl. Phys.* **33** (1962) 3580.
[163] Pollack, Freitag, and Morris, *Electrochem. Tech.* **1** (1963) 96.
[164] Geppert, *Proc. I.E.E.E.* **51** (1963) 223.
[165] Chopra, *J. Appl. Phys.* **36** (1965) 184.
[166] Beam and Armstrong, *Proc. I.E.E.E.* **52** (1964) 300.
[167] Many, Goldstein, and Grover, *Semiconductor Surfaces*, John Wiley and Sons, Inc., New York, 1965.
[168] Ridley, *Proc. Phys. Soc.* **82** (1963) 954.

Notes added in proof.

1. Our discussion of the ionic current in anodic oxide films on tantalum was mostly confined to the behavior in dilute solutions which was believed to be independent of the solute. Deviations were indeed known in various concentrated or partially non-aqueous solutions. These we believed to be due partly to the change in the activity of water, though inclusion of solute species was known to occur. It has now been known [Randall, Bernard, and Wilkinson, *Electrochem. Acta* **10** (1965) 183] that the solute may have a profound effect even in dilute solutions. Thus, in dilute phosphate solutions, phosphate is taken up by the outer part of the growing film and causes a tightening-up of the structure which gives a lower dielectric contact and ionic conductivity in this part of the film. The two-layer nature of the structures also has implications for the question of the relative mobilities of metal and oxygen ion for which the reader is referred to the original.

2. In connection with Claussen's work referred to on p. 183, the source of the change with thickness may have been in the particular way the films were sputtered. Mills, Young, and Zobel [*J. Appl. Phys.* **37** (1966) 1821] did not observe any variation in optical properties with thickness.

4

An Economic Study of the Electrochemical Industry in the United States

G. M. Wenglowski*

I. OBJECTIVES, METHODOLOGY, AND SOURCES

The object of this study is threefold:

1) To determine the size and the composition of the Electrochemical Industry in the United States.

2) To measure any changes in the size and composition of this industry during the period, 1958 to 1963.

3) To note any indications regarding the future size and composition of this industry, arising from the preceding analysis.

For the purposes of this study, the electrochemical industry is defined as the aggregate of all business entities whose major production process *or* whose final output involves electrolysis and other like phenomena accompanying the passage of a current through the solution of an electrolyte.[1] The industry, as defined, includes two major types of firms; those whose major production process is electrochemical, for example, aluminum producers, and those whose final output is electrochemical in nature, such as battery producers. This definition of the industry, however, excludes those firms whose major production process involves electric arc furnace techniques.

1. Economic Measures and Concepts

Since this report was written primarily for readers with a background in chemistry, a brief discussion of the economic measures

*Graduate student in Economics at the University of Pennsylvania.

and concepts used is in order. Two measures of size are used in this paper: shipments and value added. The shipments measure is a gross measure of economic activity, and it contains a good deal of double counting. The value added measure represents an attempt to eliminate the shortcomings of the shipments data. Value added is a better measure of the contribution of an industry to total national output.

2. Shipments

The value of shipments is closely related to that of sales. Shipments is, however, a broader concept than sales, although in most cases the two figures will be approximately equal. According to the U.S. Department of Commerce, a major source of the data in this study, "shipments . . . include all items made by or for each establishment from materials owned by it, whether sold, transferred to other plants of the same company, or shipped on consignment."[2] Large discrepancies between sales and shipments figures may arise in the case of vertically integrated firms.* Where there is vertical integration of the successive production processes within a single firm, as in the case of the integrated aluminum or copper producer who mines, smelts, refines, and fabricates, the dollar value of the shipments of all but the final product in the integrated chain will *exceed* the value of sales by the amount of interplant transfers within the same company.

In this study, shipments or estimated shipments are used as the gross measure of size. The choice of shipments as the gross measure of size was based on practical and conceptual considerations. *The Census of Manufactures*, compiled by the U.S. Department of Commerce, which contains the most detailed set of gross volume and value added data for manufactured products in the United States, uses shipments as the gross measure of size. In addition, on conceptual grounds, it is argued that the size of the electrochemical industry at any moment in time should be independent of the degree of vertical integration of its members, i.e., an ingot of primary aluminum should be counted whether it is *sold* to another firm for fabrication or *shipped* to a fabricating plant within an integrated firm. In the case of those products where

---

*A firm which is involved in the successive production processes required to produce a final output is said to be vertically integrated.

shipments data were not available, shipments were estimated by (a) sales, for the nonvertically integrated firms, or (b) production, adjusted for any change in inventories, to estimate shipments for firms of a vertically integrated nature. *Estimated* shipments measures, however, were generally used only for the smaller items in the study. Over 75% of the value of shipments in the study is based on published Department of Commerce shipment figures.

3. Value Added

The total shipments figure for the electrochemical industry over-states the contribution of this industry to the gross national product of the United States (GNP). For this reason, a second measure of size, value added, is used. The value added measure isolates the contribution of a productive activity, in this case electrochemistry, to the total output of the national economy. The shipments of the electrochemical industry overstate this contribution because: (a) the value of these shipments includes the value of goods and services which are the final outputs of other industries; (b) the value of shipments may count more than once the value of the final output of the electrochemical industry. The problem is most easily understood if the dollar value of shipments is regarded as a dollar value built up from the expenses and profits of an enterprise. The value of shipments of a firm equals the sum of the dollar values of wages and salaries, purchases of goods and services, other operating expenses, and the profit or loss of the firm. Included in these expenses, which sum to the value of a firm's shipments, are dollar outlays for the purchase of the final goods and services produced by *other* firms. The other firms may be either "members" of the electrochemical industry or of some other industry. In the latter case, the value of the firm's shipments includes the value of another industry's output. In the former case, the problem of "multiple counting" occurs.

For example, the value of shipments of the aluminum industry has embodied in it a large dollar value of purchases of electricity from firms within the electric utility industry. Thus, the value of aluminum shipments measures not only the contribution of the aluminum industry but also the contribution of a portion of the electric utility industry to the GNP. To get at the contribution of aluminum alone, the dollar value of the aluminum firms' purchases of all goods and services from other firms must be deducted from the value of shipments of the aluminum firms. The residual is the value added or contribution of aluminum to GNP.

Value Added = Shipments — Purchases of Goods and Services From Other Firms

This is a simplified definition of value added. For an exact account of the manner in which the Commerce Department obtains its value added figures see below.*

4. Multiple Counting

The use of the value added measure eliminates the contribution of other industries to GNP which is contained in the shipments data of the electrochemical industry. The use of value added figures also prevents the multiple counting of the contribution of electrochemistry to the national economy. Consider, for example, the production of batteries by a battery manufacturer and the manufacture of electrolytic manganese dioxide by a chemical firm, specializing in the production of electrolytic manganese dioxide solely for sale to the battery manufacturer. Simply summing the *shipments* of the battery and chemical firms would count the value of electrolytic manganese dioxide twice, once as electrolytic manganese dioxide, *per se*, and again in the shipments value of the batteries. The value added, computed by deducting the purchases of goods and services from the value of shipments, eliminates such multiple counting. In the value added measure, electrolytic manganese dioxide would be counted only once, as the shipments of the chemical firm, in the above example. The shipments of the battery producer would be reduced by the value of its purchases of electrolytic manganese dioxide from the chemical firm.

*"... the measure (value added) was obtained by subtracting the cost of materials, supplies and containers, fuel, purchased electric energy, and contract work from the value of shipments for products manufactured plus receipts for services rendered.

"... this measure of value added was adjusted by taking into account the following items:

(a) value added by merchandising operations (that is, the difference between the sales value and cost of merchandise sold without further manufacture, processing or assembly) plus

(b) the net change in finished goods and work-in-process inventories between the beginning and end of the year."—U.S. Department of Commerce, *1958 Census of Manufactures, Vol. II*, p. 13.

For each item in the study two measures of size are presented. The shipments measure, remember, is subject to two major shortcomings:

1) It may include the value of the final outputs of other industries as well as that of the electrochemical industry.

2) It may count more than once the contribution of the electrochemical industry to GNP.

The value added measure comes closer to reflecting the true importance of an industry in the national economy.* Both value added and shipments, except when stated otherwise, are measured in *current* 1958 or 1963 dollars, i.e., unadjusted for price changes. All figures are for United States production.

This study treats the composition of the electrochemical industry on two levels; the product composition as well as the composition by company is presented.

5. Methodology

The methodology of the study consisted of the following:

1) Determining those sectors of the national economy where electrochemistry was involved in either major production processes or final outputs.

2) Measuring the degree of electrochemistry's involvement in each sector, i.e., the percent of all production processes in a given sector that were electrochemical, or the percent of all final outputs in a sector that were electrochemical in nature.

3) Determining the *overall* size in terms of shipments and value added of each sector where some electrochemistry was present in major processes or outputs.

4) Calculating the total value of shipments and value added in each sector *due to electrochemistry*, by applying the per cent or degree of involvement of electrochemistry to the overall value of shipments and value added of the sector.

---

*"Value added avoids the duplication in the value of shipments figure which results from the use of products of some establishments as materials by others. Consequently, it is considered to be the best value measure available for comparing the relative economic importance of manufacturing among industries and geographic areas."—U.S. Department of Commerce, *1958 Census of Manufactures, Vol. II,* p. 13.

5) Summing the shipments and value added due to electrochemistry for all sectors into an industry figure and determining
the composition of this magnitude by product and by company.

6. Sources

The major sources of information used in compiling a list of those
sectors of the economy where electrochemistry was involved were:
The Encyclopedia of Electrochemistry by Clifford A. Hampel,
Electrochemical Engineering by Charles L. Mantell, *Electrochemistry*
by Giulio Milazzo, and over one hundred interviews and written
correspondence with member firms of the industry. In determining
the percent of each sector's total shipments and value added which
could be labeled as *due to electrochemistry*, the U.S. Bureau of
Mines' *Minerals Yearbook*, Standard and Poor's *Industry Survey*,
the International Tin Council's *Statistical Yearbook* were used
along with individual industry studies and the general sources
listed above. Over 75% of the shipments data in the study was
obtained from the *Census of Manufactures* and the *Current Industrial Reports* of the U.S. Department of Commerce. Two major
adjustments were made in the Department of Commerce shipments
figures for certain nonferrous metals and for certain inorganic
chemicals. These adjustments are explained in another section of
this paper. The value added figures in all cases were based on the
Census of Manufactures' data, which includes shipments and value
added figures for each four digit code of the government's Standard
Industrial Classification, (SIC). The products in the study were put
into the proper SIC four-digit categories, and the value added for
the product was estimated by applying the value added—shipments
ratio of the proper SIC category (calculated from Census of
Manufactures data) to the shipments value of the product. The
major sources for the company breakdown were *The Journal of
Metals, Minerals Yearbook, Standard & Poor's Industry Survey,
Moody's Industrial Manual, Dun and Bradstreet's Million Dollar
Directory, Oil, Paint, and Drug Reporter's Buyers' Directory,* and
the *Thomas Register*.

All figures in the study are documented and those items of a
purely estimated nature, where no published data were available,
are marked as such and the basis of the estimate noted. Personal

interviews and correspondence with representatives of the electrochemical industry played an essential role in compiling the data for those sectors of the industry where published figures were nonexistent.

II. OVERALL OBSERVATIONS ON THE ELECTROCHEMICAL INDUSTRY

In this section, the findings of the study regarding the electrochemical industry as a whole are presented. These aggregate figures were built up from the individual product and activity data according to the methodology outlined in section I. The individual product and activity figures are presented in section III. In this section, our concern is with the absolute and relative size of the industry and with the recent price movements within this industry.

1. Absolute Size

The size of the electrochemical industry in the United States in terms of value added and total shipments for 1958 and 1963 is given below in Table 1.

These figures are in the dollar values of the reporting year, unadjusted for price changes. As a result, the change in these totals between 1958 and 1963 is due to both price changes and real changes in economic activity. A comparison of the 1958 and 1963 totals in constant dollars, i.e., with prices held constant, will be presented at the end of this section, after a price index for the electrochemical industry has been constructed.

Table 1

U.S. Electrochemical Industry Shipments and Value Added in Millions of Current Dollars

	1958	1963
Total shipments	$4750.5 MM	$6431.5 MM
Value added	$2225.6 MM	$3087.0 MM

Source: Our construction—see section III.

2. Relative Size

A clearer picture of the size of the electrochemical industry in the United States is provided by a comparison of this industry with the magnitude of all U.S. manufacturing and with the size of another well-known industry. The data for all U.S. manufacturing is taken from the Commerce Department's *Census of Manufactures*. In 1958, the total shipments figure for all U.S. manufacturers was not published because of the extensive double counting involved in these figures. The value added by all U.S. manufacturers, however, was published. In 1958, the value added for all U.S. manufacturers, including administrative and auxiliary units, totaled $141,270.3 million (MM).[3] The electrochemical industry accounted for 1.58% of this total. In 1963, total shipments data for all U.S. manufacturers were published, along with a warning of the double counting which is contained in these figures. The total shipments for all manufacturers were $417,213.0 MM in 1963. Of this amount, the electrochemical industry claimed 1.54%. The value added by

Table 2

Total Shipments and Value Added of the Electrochemical Industry and all U.S. Manufacturing in Millions of Current Dollars

All U.S. Manufacturing:	1958	1963
Total shipments	N.A.	$417,213.0 MM
Value added	$141,270.3 MM	$189,995.0 MM
U.S. Electrochemical Industry:	1958	1963
Total shipments	$4750.5 MM	$6431.5 MM
Value added	$2225.6 MM	$3087.0 MM

Source: For all Manufacturing: *U.S. Census of Manufactures* 1958, 1963, U.S. Department of Commerce.
For electrochemical industry: Our construction.

all U.S. manufacturing in 1963 amounted to $189,995.0 MM.[4] The electrochemical industry accounted for 1.62% of this amount. These results are all summarized in Table 2 and Chart 1.

CHART I

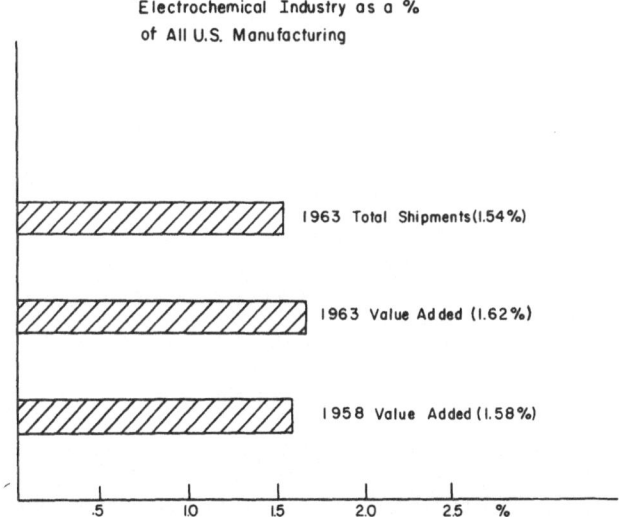

Electrochemical Industry as a %
of All U.S. Manufacturing

1963 Total Shipments(1.54%)

1963 Value Added (1.62%)

1958 Value Added (1.58%)

.5   1.0   1.5   2.0   2.5   %

We may conclude that in 1958 and 1963 the electrochemical industry contributed over $1\frac{1}{2}$ ¢ of every dollar of value added by all U.S. manufacturing, and that during the years 1958 to 1963, this contribution increased by 2.5%, reflecting a higher rate of growth for the electrochemical industry than for all U.S. manufacturing.

The size of the electrochemical industry may also be compared with the size of the industrial chemical industry. In 1964, a study of the industrial chemical industry in the United States was prepared by Jules Backman, Research Professor of Economics, New York University, and published by the Manufacturing Chemists' Association.[5]

It is important to be precise with regard to the definition of the industrial chemical industry as it is compared with the electrochemical industry. The products of the industrial chemical industry occupy the two government SIC codes, 281, inorganic and organic chemicals, and 282, plastics materials, synthetic resins, synthetic rubber, synthetic, and other man-made fibers except glass. A list of the major product categories in SIC 281 and 282 is contained in Table 3A.

In 1958, the total shipments of the industrial chemical industry

Table 3A
Composition of Industrial Chemical Industry

SIC 281:
Alkalies and chlorine
Industrial gases
Cyclic (coal tar) crudes
Dyes, dye (cyclic) intermediates and organic pigments (lakes and toners)
Inorganic pigments
Industrial organic chemicals, N.E.C.
Industrial inorganic chemicals, N.E.C.

SIC 282:
Plastics materials, synthetic resins and nonvulcanizable elastomers
Cellulosic man-made fibers
Synthetic organic fibers, except cellulosic

Source: *1958 Census of Manufactures*, U.S. Department of Commerce.

were not available. The value added by this industry in 1958, however, was $6160.0 MM, or approximately three times the contribution of the electrochemical industry in the same year.[6] In 1963, the total shipments and total value added of the industrial chemical industry were $16,515.0 MM and $9034.0 MM, respectively.[7] In each case, the electrochemical industry was approxi-

Table 3B

Total Shipments and Value Added for the Electrochemical and Industrial Chemical Industries in Current Dollars

U.S. Industrial Chemicals Industry:

	1958	1963
Total shipments	N.A.	$16,515.0 MM
Value added	$6160.0 MM	$9034.0 MM

U.S. Electrochemical Industry:

	1958	1963
Total shipments	$4750.5 MM	$6431.5 MM
Value added	$2225.6 MM	$3087.0 MM

Source: For Industrial Chemicals: *U.S. Census of Manufactures*, 1958, 1963, U.S. Department of Commerce.
For electrochemical industry: our construction.

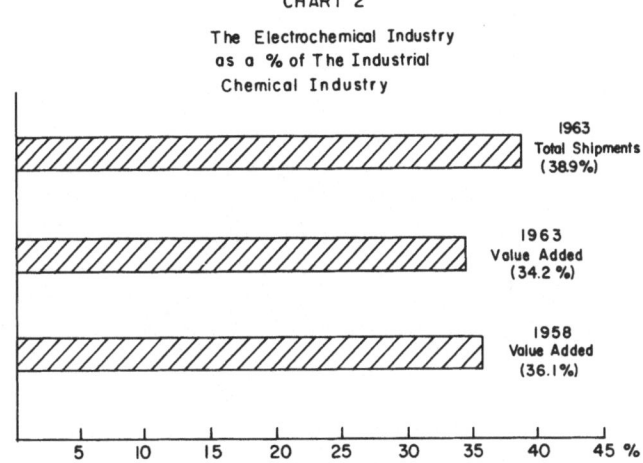

CHART 2

The Electrochemical Industry
as a % of The Industrial
Chemical Industry

1963
Total Shipments
(38.9%)

1963
Value Added
(34.2 %)

1958
Value Added
(36.1%)

mately one-third the size of the industrial chemical industry. These results are summarized in Table 3B and Chart 2.

3. Recent Price Movements in the Electrochemical Industry

There is no published price index for the electrochemical industry, as such. There do exist, however, wholesale price indices for the major components of the electrochemical industry, viz. nonferrous metals, industrial chemicals and electric machinery and equipment.[8] The coverage of these indices is broader than the coverage of the electrochemical industry within each component. An approximation of the recent price movements within the electrochemical industry, however, may be obtained from a weighted average of the price indices for these components.

To digress for a moment, a price index is a measure which shows the change in prices between any given year and a base year or base period of years. Price indices are constructed by specifying a representative set of goods or products (the composition of the set depending upon the economic sector or activity for which the index is constructed) and by noting the change in the total price of this *fixed* set of goods over time. The price of the set of goods in a selected base year (or years) is divided into the price of the same set of goods in other years to obtain the value of the price index for each of these years in terms of the base year. The value of the price

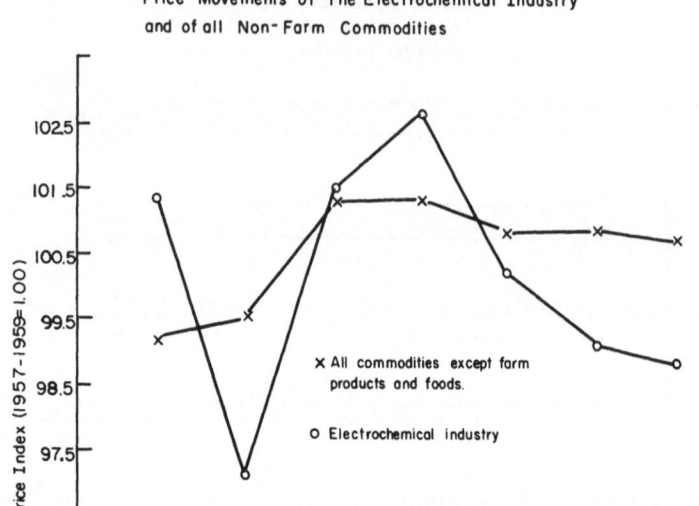

CHART 3

Price Movements of The Electrochemical Industry
and of all Non-Farm Commodities

X All commodities except farm
products and foods.

O Electrochemical industry

Source: Dept. of labor-bureau of labor statistics
Electrochemical index - our construction

index for the base year (or years) equals 1.00. An increase in prices over those prevailing in the base year yields a price index value greater than 1.00. A decrease in prices from those prevailing in the base year yields a price index value less than 1.00. Finally, to express the value of any commodity or group of commodities in terms of "constant dollars," i.e., with prices held constant at their base-year levels, divide the dollar value of the commodity by the value of the price index for the year in question. The resulting quotient is "adjusted for price changes," and comparison of such price adjusted figures reflects changes in *real* economic activity alone, unobscured by price movements.

The time period under review is 1957 to 1963. The base period of the price index which will be constructed for the electrochemical industry is 1957–1959. For the period 1957 to 1959, the wholesale price indices for the three components mentioned above are weighted in the electrochemical industry index according to the relative importance of these components in the structure of the industry during 1958. For 1960 to 1963, the weights are derived from the 1963 structure of the electrochemical industry.

The constructed electrochemical price index is presented in Chart 3, along with a price index for all commodities except farm products and foods (1957–1959 = 1.00), published by the U.S. Department of Labor.

Two major characteristics of the electrochemical industry index are apparent:

1) The index fluctuated more than the all commodities—

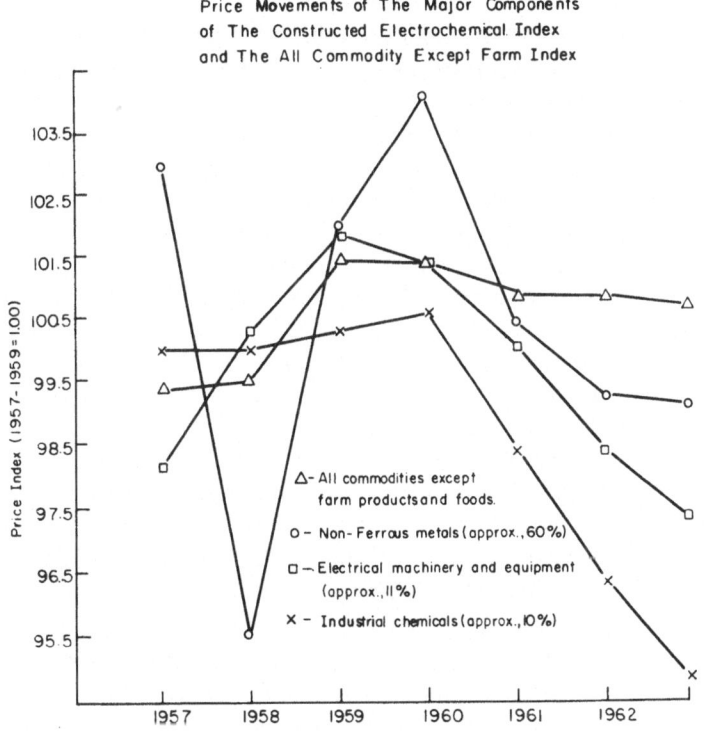

CHART 4

Price Movements of The Major Components
of The Constructed Electrochemical Index
and The All Commodity Except Farm Index

Source : Dept of Labor - Bureau of Labor Statistics
Electrochemical Index - Our Construction

except farm products and foods index—during the 1957–1963 period.

2) The index fell relatively more than the index for all commodities except farm products and foods during the most recent period. Between 1960 and 1963, the all commodity—except farm index—fell from 101.3 to 100.7, whereas the electrochemical industry index fell from 102.7 to 98.8.

Chart 4 plots the movement of each of the three major wholesale price indices, which were averaged to obtain the electrochemical industry index, along with the all commodity except farm products and foods index. Beside the label for each graph is the approximate percent weight attached to each component in the constructed electrochemical industry index. These percentages do not sum to 100, since a weight of approximately 19 % was given to the all commodity except farm index, which was averaged in the electrochemical industry index to reflect that portion of the electrochemical industry (approx. 19 %) accounted for by the many minor product classifications in this industry.

All of the components fluctuate more than the all commodity except farm index. In addition, there is a strong positive correlation between the movements of all three components of the electrochemical industry index.

Among the major components of the electrochemical industry, therefore, there appear to be no offsetting price characteristics, which would reduce the overall instability and decline of the prices in the electrochemical industry. In comparing the electrochemical industry price index with the index for industrial chemicals, the same price instability and even a greater degree of price decline—in the case of industrial chemicals—is observed. This decline in prices of industrial chemicals is generally attributed to the reduction of the price on new products as companies seek to broaden their markets.[9] To what extent the same conclusion can be applied to the electrochemical industry is left as an open question.

The change in the dollar value of total shipments and value added, unadjusted for price change, reflects both changes in real economic activity and changes in price. A comparison of the total shipments and value added of the electrochemical industry in 1958 and 1963 with prices held constant at the average 1957–1959 level reflects only the change in the economic activity of this industry.

Table 4

**U.S. Electrochemical Industry Total Shipments
and Value Added in Constant Dollars**

	1958	1963
Total shipments	$4890.4 MM	$6509.0 MM
Value added	$2291.1 MM	$3124.2 MM

Source: Our construction.

The price index for the electrochemical industry, constructed above, can be used to obtain the 1958 and 1963 shipment and value added figures in constant 1957-1959 dollars. The value of this price index for 1958 and 1963 was 97.14 and 98.81, respectively. Over the period 1958–1963, prices in the electrochemical industry have risen, although in both cases prices were below their 1957–1959 levels. Comparison of *unadjusted* dollar totals for the two years, 1958 and 1963, therefore, overstates the increase in real economic activity in the industry, part of the increase being due to rising prices. The total shipments and value added figures for the industry in constant 1957-1959 dollars is presented in Table 4. A more detailed analysis of the growth of the industry is the subject of section V.

III. THE COMPOSITION OF THE ELECTROCHEMICAL INDUSTRY—PRODUCTS

In this section, the composition of the electrochemical industry by product groupings in 1958 and 1963 is presented. First, the composition in terms of broad product types is given. This is followed by a breakdown of each of these broad categories. The section concludes with a ranking of the most important products of the electro chemical industry, taken as a whole. This section may be regarded as the basis for the totals presented in section II.

1. Composition—by Product Type

In Table 5, the seven major product categories included within the electrochemical industry are given along with the total value added for each category in 1958 and 1963 and each category's percentage

Table 5

Electrochemical Industry Composition by Product Types: 1958, 1963

	1958		1963	
Category	Value added, $MM	Percent	Value added, $MM	Percent
(1) Processes:				
Electrowinning and refining of nonferrous metals (including by-products)	$1156.9	51.98%	$1382.6	44.79%
Surface finishing	456.3	20.50	805.8	26.10
Electrochemical synthesis	276.5	12.42	374.6	12.13
Water treatment	0.1	0.01	0.2	0.01
(2) Products:				
Batteries and capacitors	254.4	11.43	413.9	13.41
Corrosion prevention	73.8	3.32	96.1	3.11
Measuring and controlling devices	7.6	0.34	13.8	0.45
Totals	$2225.6	100.00%	$3087.0	100.00%

share of the total value added by the whole electrochemical industry in the two years. In Chart 5, the same information is presented graphically.

The largest single component of the electrochemical industry was the electrowinning and electrorefining of nonferrous metals, which accounted for slightly less than one-half of the value added by the industry in 1963. The surface-finishing component was second in size, followed by batteries and capacitors, electrochemical synthesis, corrosion prevention, devices and water treatment, in that order. As shown in Table 5, four of these categories represent electrochemistry as it is involved in major production processes and three categories represent electrochemistry as the essential principle in the operation of the product sold.

The most dramatic changes in the structure of the electrochemical industry between 1958 and 1963 were the decline in importance of the electrowinning and refining component (from 51.98% to 44.79%) and the increased importance of surface finishing (from 20.50% to 26.10%). Surface finishing grew in importance, as will be shown later, primarily under the influence

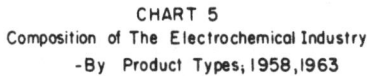

CHART 5
Composition of The Electrochemical Industry
-By Product Types; 1958,1963

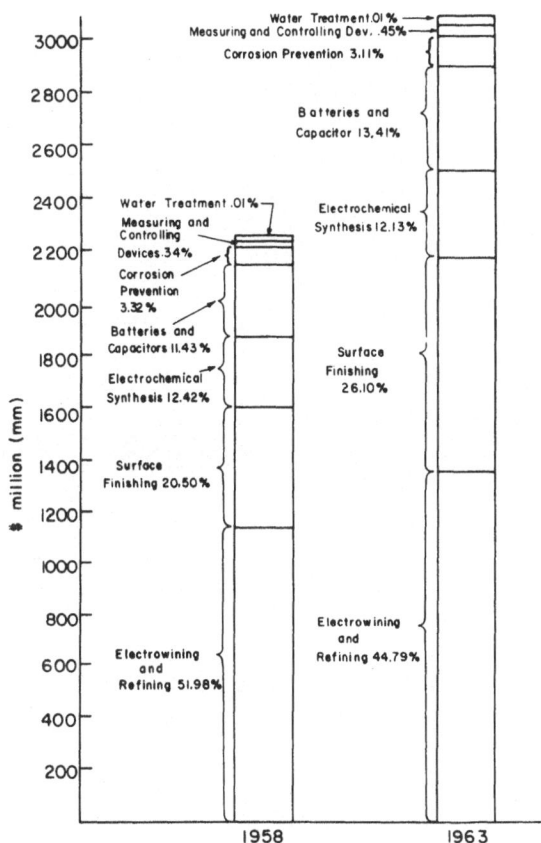

of electroplating, which received its stimulus from increased automotive production during this period. The batteries and capacitors component increased its share slightly (from 11.43 % to 13.41 %), while the four remaining components showed no significant changes.

2. Electrowinning and Electrorefining of Nonferrous Metals

This, the largest component of the electrochemical industry, accounted for slightly less than one-half of the total value added

Table 6
Electrowinning and Electrorefining of Nonferrous Metals

Product	1958				1963			
	% due to electro-chemistry	Shipments due to electro-chemistry, $MM	Value added due to electro-chemistry, $MM	Percent	% due to electro-chemistry	Shipments due to electro-chemistry, $MM	Value added due to electro-chemistry, $MM	Percent
Tin plate	91%	$904.1	$473.3	41.18%	96%	$ 950.0	$458.9	33.46%
Aluminum	100	800.0	379.2	33.00	100	1071.0	488.4	35.61
Copper	92	944.8	181.4	15.78	91	1307.4	294.2	21.46
Zinc	42	94.9	41.5	3.61	38	113.0	39.7	2.89
Gold	100	41.7	21.8	1.90	100	32.9	15.9	1.16
Magnesium	90	32.3	16.9	1.47	90	54.7	26.5	1.93
Sodium*	100	17.3	9.2	0.80	100	25.1	13.6	0.99
Silver	100	13.7	7.2	0.62	100	15.0	7.2	0.52
Copper powder†	50	8.0	5.2	0.45	50	13.2	8.3	0.60
Electrolytic manganese†	100	7.6	4.0	0.35	100	9.2	4.4	0.32
Tantalum†	40	6.0	3.1	0.27	40	10.2	4.9	0.36
Cadmium	40	4.8	2.5	0.22	40	8.7	4.2	0.31
Electrolytic manganese dioxide†	100	3.0	1.6	0.14	100	4.0	1.9	0.14

Lead	8	10.2	1.1	0.10	10	10.4	1.1	0.08
Selenium	100	5.3	1.0	0.09	100	3.5	0.8	0.06
Lithium†	100	0.3	0.2	0.02	100	0.9	0.5	0.04
Electrolytic tin	0	0	0		100	4.3	0.9	0.07
Beryllium								
Titanium								
Zirconium		Electrochemical production in laboratory quantities only.						
Vanadium								
Tungsten								
Uranium								
Rare earth metals								
Tellurium								
Cerium		Insignificant electrochemical production.						
Elemental boron								
Boron$_{10}$								
Totals		$2894.0	$1149.2	100.00%		$3633.5	$1371.4	100.00%

*Estimated for 1963.
†Estimated for 1958 and 1963.

by the industry in 1963. This category includes all primary and secondary refining or winning of metals, where the process is electrochemical in nature. In four cases, aluminum, lead, gold, and silver, only primary production was considered in this study. To the best of our knowledge, no significant amount of the secondary recovery of these four metals was done electrochemically.

Table 6 contains a list of the specific metals included, along with the percentage, in columns 2 and 6, of the total shipments and the total value added of each metal which resulted from electrochemical methods in 1958 and 1963. Table 6 also includes, for 1958 and 1963, the dollar value of shipments and value added of each metal which was *due to electrochemical processes*, and each metal's percentage share of the value added of the whole nonferrous metals category.

The percentage of the total shipments of each metal which was due to electrochemical processes was based on the general sources for electrochemistry and electrochemical engineering and on Bureau of Mines production figures, which categorize the output or the production capacity of the more important metals in terms of electrolytic and other production processes.* These percentages for electrochemical production methods were then applied to the total shipments and total value added for each metal to obtain the total shipments and value added *due to electrochemistry*, the entries in columns 3, 4, 7, and 8 of Table 6. On the basis of the value added due to electrochemistry for each metal, the percentage importance of each metal in the nonferrous metals category was determined and given in columns 5 and 9 of Table 6.

In the case of the nonferrous metals one adjustment was made to the shipments figures that were published in the Commerce Department's *Census of Manufactures*. The shipments figures given in the *Census of Manufactures* include the value of work done on toll, (i.e., the processing of materials owned by others), in the industry of the firm for which the work is done and *not* in the industry where the service is actually performed. In the case of certain nonferrous metals, this procedure caused a large discrepancy between the production in tons and shipments in tons figures

---

*Examples of these "general sources" are *Electrochemical Engineering*, by Mantell (McGraw-Hill, 1960) and *The Encyclopedia of Electrochemistry*, by Hampel (Reinhold, 1964).

included in the *Census of Manufactures*. The discrepancy was due to the fact that production or refining of materials owned by others, i.e., on toll, is not counted in the shipments of the industry where the production is accomplished, but rather is counted for the industry that *owns* the materials. This discrepancy between production and shipments due to large amounts of toll work was important in the case of copper, zinc, and lead refiners. In those cases, an adjustment was made to include the toll work in the total shipments value of the metal. The value added and shipments due to electrochemistry were then derived from these adjusted shipments figures.

In six cases, published data on the percentage of electrochemical involvement, total shipments, or value added were not available. Estimates based on correspondence with firms producing the metals in question were used in these cases. The estimated figures are noted in Table 6, and they account for a very small share of the nonferrous metals category.

One other peculiarity of Table 6 should be noted. That is, the very low value added–shipments ratio for copper. In both 1958 and 1963 the total shipments of copper were the largest of all the nonferrous metals. However, in terms of value added, copper was third in size. This reflects a relatively high ratio of purchases to shipments for copper producers.

Table 6 contains all those metals where electrochemical techniques were employed in the refining or winning process. The importance of the electrochemical process relative to other competing processes varies. In some cases, the electrochemical process was only on an experimental basis. In others, the value of shipments was insignificant, i.e., less than $1 MM.

Between 1958 and 1963, there was only one significant change in the importance of an electrochemical process *vis-à-vis* competing processes in the nonferrous metals group. That was the case of tin plating, which accounted for over $\frac{1}{3}$ of the nonferrous metals component. The percentage of all tin plate that was electrolytic increased from 91% in 1958 to 96% in 1963. This change reflects changes in U.S. tinplating which have been occurring for some time. In 1950 only 40% of all tinplate was produced electrolytically. In 1953 the portion had increased to approximately 71%, and in 1955, approximately 80% of all tinplate was electrolytically produced.[10]

CHART 6

The Composition of The Electrowinning and
Refining Component of The Electrochemical
Industry

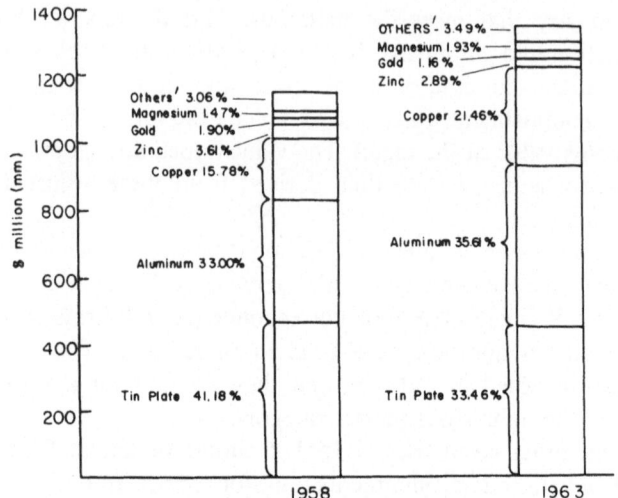

1958   1963

Sodium, Sliver, Copper Powder, Electrolytic Manganese Dioxide,
Tantalum, Cadmium, Lead, Selenium, Lithium, Electrolytic Tin,
Electrolytic Manganese.

Chart 6 depicts the change in the structure of the electrowinning and electrorefining of nonferrous metals between 1958 and 1963. The most important changes in the structure of this component were a decline in the relative importance of tinplating, a moderate increase in the share of aluminum, and a large increase in the importance of copper.

3. By-Product Metals

Electrochemical processes, primarily in the refining of copper, lead, and zinc, yield certain rare metals as by-products. The value of these by-product metals should be included within the electrochemical industry. In Table 7, the major metals derived as by-products of electrorefining operations are listed. The total value of shipments for each of these metals was based on data contained in the Bureau of Mines' *Minerals Yearbook*.[11] Since copper refining represented the major source of these by-product metals (approx. 90% of all by-product gold and 60% of all by-product silver is derived from copper refining) the value added–shipments ratio for copper was used to obtain the value added figures in Table 7. As seen in Table 7, gold and silver are by far the most important by-products of electrorefining methods. In addition, some platinum, paladium, and nickel is derived. In 1958, the shipments value of all these by-product metals totaled $40 MM, and in 1963 the figure was $50 MM.

4. Surface Finishing

The surface-finishing component of the electrochemical industry accounted for 20% of the industry in 1958 and 26% of the industry in 1963. It was the second largest component of the electrochemical industry.

Surface finishing consists of electroplating and a small amount of detinning. Tin plate is included within the nonferrous metals component and, therefore, *not* included here. In Table 8, the shipments and value added figures for these processes in 1958 and 1963 are given. The shipments value for detinning is based on data from the International Tin Council's *Statistical Yearbook*.

In the case of electroplating, a very important component of the electrochemical industry, the data was very scarce. The *Census of Manufactures*' figures include only the plating done by job shops. Use of these figures would ignore the significant amount of electroplating performed by the major automobile and appliance manufacturers. The National Association of Metal Finishers has conducted an industry survey, but once again, only the job shops were included in the results.[12]

In this study, the task of estimating the dollar value of electroplating was divided into an estimate of automotive plating and an estimate for nonautomotive plating. First, an estimate of the dollar value of electroplating per car was made. In 1950, *Plating* magazine placed this figure at about $50 per car.[13] An interview with Dr. William Blum of Washington, D.C., a noted expert in the electroplating field, placed the current estimate in the range of $50 to $100 per car. An estimate of $80 per car for automotive electroplating was used in this study. Based on the $4\frac{1}{4}$ million cars produced in the U.S. in 1958, our estimate of automotive plating was

Table 7

By-Products of Electrochemical Processes

Product	1958				1963			
	% due to electro-chemistry	Shipments due to electro-chemistry, $MM	Value added due to electro-chemistry, $MM	Percent	% due to electro-chemistry	Shipments due to electro-chemistry, $MM	Value added due to electro-chemistry, $MM	Percent
Gold	100%	$19.9	$3.8	49.35	100%	$18.5	$4.2	37.50%
Silver	100	19.6	3.8	49.35	100	29.8	6.7	59.82
Platinum		insignificant				insignificant		
Palladium		insignificant			100	0.6	0.1	0.89
Nickel	100	0.8	0.1	1.30	100	1.1	0.2	1.79
Totals		$40.3	$7.7	100.00%		$50.0	$11.2	100.00%

Table 8

Surface Finishing

| Product | 1958 | | | 1963 | | |
	% due to electro-chemistry	Shipments due to electro-chemistry, $MM	Value added due to electro-chemistry, $MM	% due to electro-chemistry	Shipments due to electro-chemistry, $MM	Value added due to electro-chemistry, $MM
Detinning*	100%	$ 6.6	$ 3.4	100%	$ 7.7	$ 3.7
Plating*	100	640.6	452.9	100	1111.0	802.1
Totals		$647.2	$456.3		$1118.7	$805.8

*Estimate for 1958 and 1963.

Table 9
Organic and Inorganic Industrial Chemicals Produced Electrochemically

Product	1958				1963			
	% due to electro-chemistry	Shipments due to electro-chemistry, $MM	Value added due to electro-chemistry, $MM	Percent	% due to electro-chemistry	Shipments due to electro-chemistry, $MM	Value added due to electro-chemistry, $MM	Percent
Sodium hydroxide	91%	$202.1	$122.7	44.38%	93%	$255.6	$153.8	41.06%
Chlorine	96	198.9	120.8	43.69	98	300.3	180.7	48.24
Hydrogen peroxide	100	22.5	12.0	4.34	15	5.6	3.1	0.83
Other alkalies	95	16.5	10.0	3.62	95	21.7	13.0	3.47
Sodium chlorate	100	8.7	4.6	1.66	100	27.9	15.2	4.06
Electrolytic hydrogen	100	6.8	4.3	1.55	100	10.1	6.2	1.65
Electrolytic oxygen	100	2.8	1.8	0.65	100	2.5	1.5	0.40

Product								
Potassium chlorate*	100	0.5	0.3	0.11	100	0.5	0.3	0.08
Ammonium perchlorate†	100	0	0		100	1.5	0.8	0.21
Dialdehyde starch								
Tetraethyl lead								
Tetramethyl lead								
Adiponitrile								
Fluorocarbons								
Potassium perchlorate	} Insignificant electrochemical production							
Sodium perchlorate								
Potassium bromate								
Sodium bromate								
Periodates								
Nitrogen trifluoride								
Totals		$458.8	$276.5	100.00%	100	$625.7	$374.6	100.00%

*Estimated for 1958 and 1963.
†Estimated for 1963.

$340.6 MM. In 1963, our estimate of the automotive plating for the over $7\frac{1}{2}$ million cars produced was $611.0 MM.

In the case of the myriad of nonautomotive products which involve electroplating (plumbing fixtures, table ware, household appliances, etc.), our estimate was $300 MM in 1958 and $500 MM in 1963. No published figures were available for the value of non-automotive plating. The estimates given above were based on an interview with Dr. Blum and other correspondence, where the general feeling was that nonautomotive plating was almost equal in value to automotive plating. The total figure for electroplating shipments and value added, automotive and nonautomotive (excluding tin plate), is given in Table 8. The value added figure was obtained in the usual manner, by applying the value added–shipments ratio for electroplating from the *Census of Manufactures* to the value of shipments.

5. Electrochemical Synthesis

This component of the electrochemical industry was third in size in 1958 and fourth in 1963. This category includes all those in-organic and organic industrial chemicals which are produced electrochemically. In Table 9, a list of these chemicals is given. The percent of the total shipments of each chemical which was due to electrochemical processes is found in columns 2 and 6. These percentages were derived from the general sources in the field of electrochemistry and electrochemical engineering. In the case of chlorine and the alkalies, the percentages were obtained from a recent Master's Thesis in Business Administration on the Chlor-Alkali Industry, in which the productive capacities for each product are categorized by process.[14] The shipments figures in all cases except two, potassium chlorate and ammonium perchlorate were obtained from the *Census of Manufactures*. The figures for potassium chlorate and ammonium perchlorate are estimates based on correspondence and interviews.

An adjustment was made to the shipments data for chlorine and sodium hydroxide as contained in the *Census of Manufactures* to allow for the large amount of captive production, i.e., production for the internal use of the producer, which takes place in the chlor-alkali industry. The amount of the captive production is reflected in an excess of the production figures in tons over the shipments figures in tons. This discrepancy was important in the case of chlorine and sodium hydroxide. In these two cases an adjustment was made to increase the shipments figure by the amount of the captive production.

From Table 9, it is apparent that electrochemical processes are concentrated mainly in the production of inorganic chemicals. There have been some recent successes in electrochemical processes for organics, but in 1963 most of this production was insignificant (less than $1 MM). After 1963, Nalco Chemical's plant for the production of tetraethyl and tetramethyl lead, using an electrochemical process, was back in operation, following a fire in February of 1963.[15] Miles Chemical is currently using an electrochemical process for producing dialdehyde starch. However, 1963 shipments were estimated at less than $1 MM. Perhaps, the largest application of electrochemical processes in the area of organic chemicals will stem from Monsanto Chemical's decision to build a large facility for the production of adiponitrile. Electrochemical organics, therefore, although termed insignificant in 1963, have recently been the subject of much interest and activity.

In the case of the inorganic industrial chemicals, by far the most important products were chlorine and sodium hydroxide, which together in 1963 claimed just under 90% of the value added by all inorganic and organic chemicals produced electrochemically. Hydrogen peroxide was second to the chlor-alkali group in 1958, although by 1963 the introduction of a new non-electrochemical process had severely reduced its importance. Electrolytic hydrogen, oxygen, sodium chlorate, potassium chlorate, and ammonium perchlorate all accounted for a very small share of those industrial organic and inorganic chemicals produced electrochemically.

Chart 7 depicts the change in the structure of this segment of the electrochemical industry between 1958 and 1963. The most dramatic change was the decline in importance of the electrochemical process for producing hydrogen peroxide. In 1958, the electrochemical process was the sole process for producing hydrogen peroxide. By 1963, this process accounted for only about 15% of all production. Sodium chlorate grew in importance during this period and ammonium perchlorate, used largely in rocket fuels, grew from insignificant 1958 shipments to a size of $1½ MM

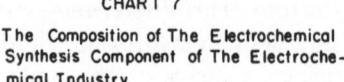

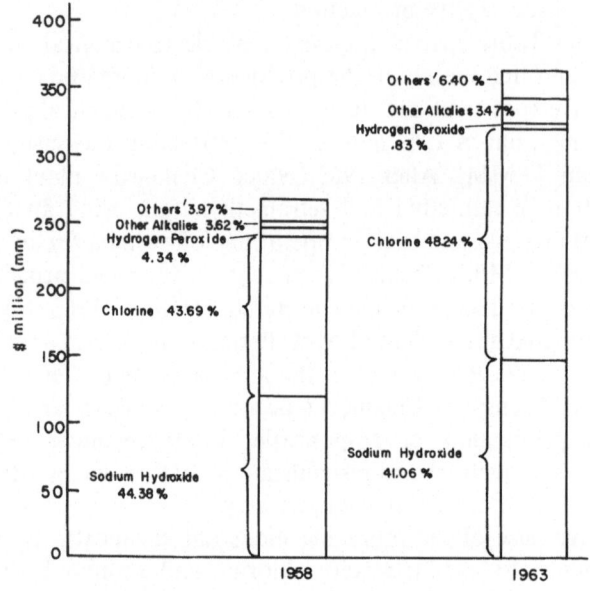

CHART 7

The Composition of The Electrochemical Synthesis Component of The Electrochemical Industry

Sodium Chlorate, Electrolytic Hydrogen, Electrolytic Oxygen, Potassium Chlorate, Ammonium Perchlorate

in 1963. The chlor-alkali group maintained its share of the electrochemical-synthesis component at slightly less than 90%.

6. Batteries and Capacitors

In 1963, the value added by batteries and capacitors accounted for over 13% of the value added by the entire electrochemical industry. Table 10 presents a breakdown of these products. Since the operation of these devices is based on electrochemical principles, 100% of their shipments and value added is recorded as part of the electrochemical industry. The 1958 shipments and value added figures were obtained from the 1958 *Census of Manufactures*. The 1963 data for aluminum and tantalum capacitors are preliminary Department of Commerce estimates, and the 1963 figure for AC-electrolytic capacitors is our estimate. The remaining 1963 data were obtained from the 1963 *Census of Manufactures*.

Table 10
Batteries and Capacitors

Product	1958				1963			
	% due to electrochemistry	Shipments due to electrochemistry, $MM	Value added due to electrochemistry, $MM	Percent	% due to electrochemistry	Shipments due to electrochemistry, $MM	Value added due to electrochemistry, $MM	Percent
Storage batteries	100%	$361.5	$147.1	57.82%	100%	$476.8	$226.5	54.72
Primary batteries (dry and wet)	100	135.7	74.1	29.13	100	190.2	110.9	26.80
Electrolytic capacitors (AC)*	100	2.0	1.2	0.47	100	2.5	1.5	0.36
Electrolytic capacitors (aluminum)*	100	35.3	22.3	8.77	100	62.4	39.4	9.52
Electrolytic capacitors (tantalum)*	100	15.3	9.7	3.81	100	56.4	35.6	8.60
Totals		$549.8	$254.4	100.00%		$788.3	$413.9	100.00%

*Estimated for 1963.

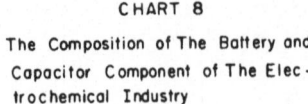

CHART 8

The Composition of The Battery and
Capacitor Component of The Electrochemical Industry

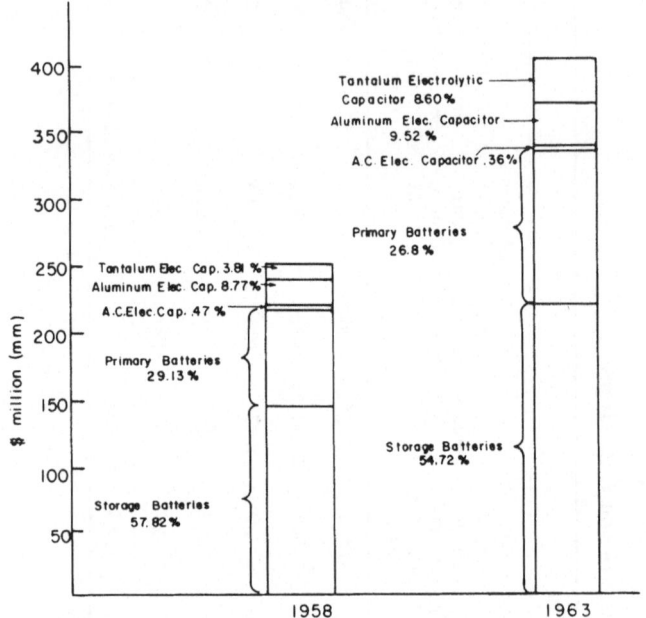

In both 1958 and 1963, batteries accounted for over 80%
of the value added in this component of the electrochemical industry, with the residual (approx. 20%) due to electrolytic capacitors.
In Chart 8, the changing structure of this segment of the industry
is depicted. There has been about a 5% decline in the share of
batteries and a parallel increase in the importance of electrolytic
capacitors, due largely to the increase in shipments of tantalum
capacitors.

In 1958, the value of shipments of tantalum capacitors was
only $15.3 MM and the value added only 3.8% of that added by the
entire batteries and capacitors component. By 1963, the value of
tantalum capacitor shipments had risen to $56.4 MM, and their
share had become 8.6%.

7. Water Treatment

Water treatment is the smallest component of the electrochemical industry. The main activity of the industry in the area of water treatment is desalination. In 1958 and 1963, desalination research involving electrochemical techniques was carried on by the U.S. Office of Saline Water. The costs involved in this research, however, were well below $1 MM and, therefore, rated as insignificant in Table 11. In addition, in 1958 and 1963 there were approximately 5 and 25 plants, respectively, treating brackish water by electrodialysis in U.S. municipalities.

The principal developer of these plants for the U.S. and foreign market is Ionics Inc. of Cambridge, Massachusetts. The estimated operating expenses of these electrodialysis plants in the U.S. in 1958 and 1963 were set at $100,000, and $400,000, respectively, by Mr. Russell L. Haden, Jr., President of Ionics, Inc. Since no shipment values exist for the treated water, the value of shipments arising from electrodialysis was estimated by the operating costs of the plants. The total value of electrochemical water treatment, therefore, was estimated at $.1 MM in 1958 and $.4 MM in 1963. Value added was estimated by using a 50% value added–shipments ratio.* Table 11 summarizes the data for this section.

The two remaining components of the electrochemical industry are corrosion prevention and controlling and measuring devices. In both of these components, our interest is only in those devices whose operation is based on electrochemical principles. Such items fall within our definition of the electrochemical industry. There were no published figures available for the shipments value of either of these two components in 1958 or 1963. As a result, estimates based on interviews and correspondence with members of the industry are given.

8. Controlling and Measuring Devices

As mentioned above, this category includes all those measuring and controlling devices whose operation is based on electrochemical principles. The shipments figures for this component of the electrochemical industry are the result of a mail survey of the major

*There was no value added or shipments data for electrodialysis water treatment in the *Census of Manufactures*. Fifty percent, which represents an average value added–shipments ratio, was used to estimate value added by electrodialysis.

Table 11

Water Treatment

Process	1958			1963		
	% due to electro-chemistry	Costs due to electro-chemistry, $MM	Value added due to electro-chemistry, $MM	% due to electro-chemistry	Costs due to electro-chemistry, $MM	Value added due to electro-chemistry, $MM
Desalination research (Office Saline Water) Electrodialysis*	100%	Insignificant $0.1	$0.1	100%	Insignificant $0.4	$0.2
Totals		$0.1	$0.1		$0.4	$0.2

*Estimated for 1958 and 1963.

Table 12

The Major Measuring and Controlling Devices Which Operate Electrochemically

Product	1958			1963		
	% due to electro-chemistry	Shipments due to electro-chemistry, $MM	Value added due to electro-chemistry, $MM	% due to electro-chemistry	Shipments due to electro-chemistry, $MM	Value added due to electro-chemistry, $MM
P.H. meters*	100%	$1.6	$1.0	100%	$2.0	$1.3
Electrolytic analyzers*	100	0.4	0.2	100	0.5	0.3
Polarographs*	100	1.0	0.6	100	1.5	0.9
Glass and specific ion electrodes*	100	5.0	3.0	100	10.0	6.3
Oxygen electrode*	100	0.1	0.1	100	0.9	0.6
Electrolytic hygrometer*	100	0.5	0.3	100	1.1	0.7
Conductivity meters*	100	2.2	1.3	100	3.2	2.0
Titrators*	100	1.0	0.6	100	1.8	1.1
Electrochemical integrators*	100	0.8	0.5	100	1.0	0.6
Totals		$12.6	$7.6		$22.0	$13.8

*Estimated for 1958 and 1963.

producers of these devices. We asked the correspondents to add to or delete from our tentative list of devices, embodying electrochemical principles, and to estimate, on the basis of their experience, the U.S. shipments or sales of these devices. Twelve questionnaires were sent out and eight returned. The results are presented in Table 12. The value-added figures in Table 12 were obtained by applying the value added–shipments ratio for SIC group 3811, scientific instruments, published in the *Census of Manufactures*, to the estimated shipments figure.

The value added by the major devices in this group in 1958 and 1963 is estimated at $7.6 MM and $13.8 MM, respectively. Their share of the total electrochemical industry was less than $\frac{1}{2}\%$ in both 1958 and 1963. Some devices which operate electrochemically may well have been omitted from Table 12. The major devices of this nature, however, have been included, and any omissions will not be of sufficient magnitude to affect the overall results of this study.

9. Corrosion Prevention

Corrosion prevention is another area where accurate published data were unavailable. In 1958 and 1963, the share of this component in the total electrochemical industry was approximately 3%. Our estimate of the value of shipments of corrosion prevention products, which function electrochemically, was $147.7 MM in 1958 and $192.9 MM in 1963. The corrosion prevention products which were considered in this study were zinc coatings and cathodic protection systems. The estimates of the 1958 and 1963 shipments of zinc coatings were made by a representative of a leading firm in the zinc coating market. This estimate for 1958 was $3.0 MM and for 1963 $8.0 MM. These figures only include the cost of the materials applied. The cost of application, which may equal many times the cost of materials, is excluded. The value added by zinc coatings was obtained from the value added–shipments ratio of the appropriate SIC category in the *Census of Manufactures*.

In the case of cathodic protection a useful source of information was a study conducted by the National Association of Corrosion Engineers (NACE), concerning the dollar value of the influence of NACE members on the purchases of specific materials protection equipment in 1962.[16] The figure for cathodic protection in the NACE study was $210.0 MM.

An estimate of cathodic protection shipments was also made by Mr. Robert J. Walton and his staff at Walton Associates. Their estimate for 1963, broken down by users, is:

Elevated water tanks	$ 5.0 MM
Pipe lines	50.0 MM
Off-shore platforms	10.0 MM
Power plants and refineries	10.0 MM
Ships	5.0 MM
Piers	5.0 MM
Hot water tanks	10.0 MM
Exports	50.0 MM
Total	$145.0 MM

Our final estimate was approximately $145.0 MM in 1958 and $185.0 MM in 1963, for U.S. cathodic protection equipment shipments. To this figure we applied a 50% value added–shipments ratio.* Table 13 summarizes these results.

10. The Electrochemical Industry's Most Important Products

In Table 14, the major products of the electrochemical industry in terms of 1963 value added are ranked by size. Value added was chosen as the basis for this ranking because it reflects more accurately the contribution of the electrochemical industry to GNP. The value of shipments contains the value of the products of other industries as well as the electrochemical industry, and it may count more than once the contribution of the electrochemical industry. Value added, therefore, is regarded as the better index of relative importance.

The value added–shipments ratios do vary among these fourteen products, as can be seen in Table 14. Use of shipments as the basis for the ranking would change the rank.

The total value added by these major products of the electro-chemical industry was $2980.4 MM or 96.5% of the total value

*There were no value added or shipments data for cathodic protection in the Census of Manufactures. Fifty percent, which represents an average value added–shipments ratio, was used to estimate value added by cathodic protection.

Table 13
Corrosion Prevention

Product	1958			1963		
	% due to electro-chemistry	Shipments due to electro-chemistry, $MM	Value added due to electro-chemistry, $MM	% due to electro-chemistry	Shipments due to electro-chemistry, $MM	Value added due to electro-chemistry, $MM
Zinc coatings*	100%	$ 3.0	$ 1.3	100%	$ 8.0	$ 3.5
Cathodic protection*	100	144.7	72.5	100	184.9	92.6
Totals		$147.7	$73.8		$192.9	$96.1

*Estimates for both 1958 and 1963.

Table 14

The Most Important Products of the Electrochemical Industry in 1963

Rank	Description	1963 value added due to electro-chemistry, $MM	1963 shipments due to electro-chemistry, $MM	% of total electrochemical industry's 1963 value added
1	Electroplating	$802.1	$1111.0	26.0%
2	Aluminum	488.4	1071.0	15.8
3	Tin plate	458.9	950.0	14.9
4	Batteries (primary and storage)	337.4	667.0	10.9
5	Copper	294.2	1307.4	9.5
6	Chlorine	180.7	300.3	5.8
7	Sodium hydroxide	153.8	255.6	5.0
8	Cathodic protection	92.6	184.9	3.0
9	Zinc	39.7	113.0	1.3
10	Aluminum capacitors	39.4	62.4	1.3
11	Tantalum capacitors	35.6	56.4	1.1
12	Magnesium	26.5	54.7	0.9
13	Gold (excluding by-products)	15.9	32.9	0.5
14	Sodium chlorate	15.2	27.9	0.5
	Total	$2980.4	$6194.5	96.5%

added by the entire industry in 1963. The 59 remaining products of the electrochemical industry included in this study accounted for only 3.5% of the value added by the whole electrochemical industry.

The most important product of the electrochemical industry in terms of value added was electroplating. The value added by electroplating was almost 75% greater than that added by the second ranked product, aluminum. Electroplating's predominance was due to a high volume of shipments coupled with a value added–shipments ratio of over 70%.* Electroplating, itself, accounts for

*Value added–shipments = 253.7 MM/359.1 MM, in 1957. Value added–shipments = 375.0 MM/519.0 MM, in 1963—1958 *Census of Manufactures*, 1963 *Census of Manufactures*, Preliminary Report.

over $\frac{1}{4}$ of the value added by the entire electrochemical industry. If the value added by tin plate is included with the value added by electroplating, the resultant sum represents over 40% of the value added by the entire electrochemical industry.

Aluminum is ranked second with value added equal to $488.4 MM or 15.8% of the value added by the entire electrochemical industry. In third place is tin plate with a value added of $458.9 MM, 14.9% of that for the entire electrochemical industry. Batteries accounted for 10.9% of the industry, and they are fourth in rank. Copper is ranked fifth. The shipments of copper totaled $1307.4 MM, the largest of any product. But a very low value added–shipments ratio (approx. 22%) reduced it to fourth place in terms of value added. The rank of the remaining products is presented in Table 14, along with the value added and total shipments due to electrochemistry for each product in 1963 and the percentage of the total value added by the entire electrochemical industry accounted for by each product in 1963.

11. A Note on Reliability, Comprehensiveness

In any study of this nature, which gives specific figures and conclusions based on them, the reader may be concerned with regard to the figures presented and the resulting conclusions. In this paper, we have been very careful to label estimates as estimates. The estimates were used only when no published data were available. They were based on numerous interviews and correspondence. At most, however, they indicate an order of magnitude, an approximation. The role of estimates was not important in terms of the dollar value involved. In terms of the value of shipments of the electrochemical industry in 1958, only 17% of this value as presented in this study was based on estimates. In 1963, estimates accounted for only 23% of the total value of shipments presented. The remaining, and by far, the major portion of the data presented was obtained from published figures, whose sources have been given throughout this presentation.

In regard to the coverage of this study, it includes some seventy-four products produced by over 125 companies in at least seven distinct industries. But this represents only that part of the widespread electrochemical industry which we were able to quantify and catalog, given the restrictions of time and finances.*

Even at the conclusion of this economic study, a list of areas existed, where electrochemistry was known to be involved, but which had not been analyzed or quantified as to the degree or value of involvement. These areas include electroforming, electrocleaning, the etching of transistors, plating as practiced by the printing industry, electrochemistry involved in the production of the masters for phonograph records, xerography, and anodizing. These all pose unusually difficult data problems and problems in estimating the value which should be attributed to the electrochemical industry. No doubt, there are other areas which could be included in the above list.

In the two concluding sections of this paper, the position of the major firms in the industry is examined and a section on growth experience and growth prospects is presented.

IV. THE COMPOSITION OF THE ELECTROCHEMICAL INDUSTRY—COMPANIES

The major companies in the U.S. electrochemical industry are listed in this section. First, an attempt is made to rank the twenty-five most important firms on the basis of each company's value added, *due to electrochemistry*, in 1963. The value added *due to electrochemistry* was estimated as follows:

1) The major companies within each process or product category of section III were determined from the published sources outlined in section I and from interviews with representatives of companies known to be in a given category.

2) An estimate of each of these major company's percentage share of the total value added of a particular product or process category for 1963 was determined by consulting published production capacity or sales figures for the companies. Figures on the 1963 production capacity of individual firms, categorized by process

---

*Figures on the cost of the electrochemical operations performed by the Atomic Energy Commission in the area of isotope production and reduction arrived too late for inclusion in the main body of this report. The approximate costs of the combined electrochemical operations performed by the A.E.C. to produce elemental boron, lithium, fluorine, etc. were $5.7 MM in 1958 and $1.8 MM in 1963. These figures were provided by Mr. F. P. Baranowski, Director Division of Production, A.E.C.

(electrolytic, etc.), were available for the major nonferrous metals and inorganic chemicals included in section III. In addition, published industry studies and industry interviews were used.*

3) The estimate of each company's percentage share of the value added of a category was then applied to the total value added for the product or process category in section III to obtain the estimate of the value added for each company. When companies were involved in more than one category, the value added in each category by the company was assumed to obtain the company figure.

The results of this procedure are given in Table 15. For each company, the products which it produces that are included within the electrochemical industry are given, along with the value added for these products and the percent of the total value added of the entire electrochemical industry accounted for by the company's "electrochemical products."†

In 1963, the largest company in the electrochemical industry, in terms of value added due to electrochemistry, was the world's largest corporation, General Motors. The value added by the electroplating and battery manufacture carried on by G.M. in 1963 is estimated at $285.2 MM, 9.2% of the total value added by the electrochemical industry and almost 75% larger than the value added by the second-ranked company. The Aluminum Company of America, the largest U.S. aluminum producer, was second in importance in the electrochemical industry in 1963. Its value added was estimated at $166.0 MM, approximately 5% of the value added by the entire electrochemical industry. The largest U.S. tinplater, United States Steel, was the third ranked company in the electrochemical industry. The value added of U.S. Steel was approximately 5% of the value added by the entire electrochemical industry in 1963. Figures for the remaining twenty-two leading firms are given in Table 15.

The total value added of all twenty-five leading firms accounted for almost 60% of the value added by the entire electrochemical industry in 1963.

*Standard and Poor's *Industry Studies.* Interviews played an important role in determining the major firms in the batteries and capacitors, measuring and controlling devices, and corrosion prevention components.
†"Electrochemical Products," as usual, means those products produced using electrochemical processes and (or) those products sold whose operation is based on electrochemical principles.

Table 15

The Twenty-five Leading U.S. Electrochemical Firms in 1963

Rank	Company	Products due to electrochemistry	Value added of products due to electrochemistry, $MM	% of total electrochemical industry 1963 value added
1	General Motors	Electroplating, storage batteries	$285.2	9.2 %
2	Aluminum Co. of America	Aluminum	166.0	5.4
3	U.S. Steel	Tin plate	157.3	5.1
4	Reynolds Metals	Aluminum	136.3	4.4
5	Dow Chemical	Chlorine, sodium hydroxide, magnesium	128.8	4.2
6	Kaiser Aluminum	Aluminum, chlorine, sodium hydroxide	120.9	3.9
7	Anaconda Co.	Copper, zinc, gold, silver, by-products, cadmium, selenium, aluminum	91.3	3.0
8	American Smelting and Refining	Copper, zinc, gold, silver, by-products, cadmium, selenium	90.0	2.9
9	Ford Motor Co.	Electroplating, storage batteries	89.8	2.9
10	National Steel	Tin plate, detinning	83.6	2.7
11	Bethlehem Steel	Tin plate	74.4	2.4
12	Phelps Dodge	Copper, by-products	71.7	2.3
13	Electric Storage Battery	Storage batteries, primary batteries	66.2	2.1
14	Kennecott Copper	Copper, by-products, selenium	62.0	2.0
15	Pittsburgh Plate Glass	Chlorine, sodium hydroxide, sodium chlorate	47.3	1.5
16	American Metal Climax	Copper, copper powder, gold, silver, by-products, selenium	35.4	1.1
17	Ormet Corp.	Aluminum	35.0	1.1
18	Gould National	Storage batteries	34.0	1.1
19	Youngstown Sheet and Tube	Tin plate	29.8	1.0
20	Diamond Alkali	Chlorine, sodium hydroxide	27.9	0.9
21	Union Carbide	Primary batteries, electrolytic manganese	27.7	0.9
22	Globe Union	Storage batteries	27.2	0.9
23	Jones and Laughlin Steel	Tin plate	26.9	0.9
24	Kaiser Steel	Tin plate	24.2	0.8
25	Allied Chemical	Chlorine, sodium hydroxide	22.1	0.7
	Totals		$1824.7	59.1 %

In the remainder of this section a listing of the major companies in each product and process category within the electrochemical industry is given. Over 125 companies are listed. For the more important products and processes, the importance of each firm is indicated by its approximate percentage share (in parentheses) of the productive capacity or total sales in the given category.

Table 16

Major Firms in Product Areas of the Electrochemical Industry

Aluminum
Aluminum Co. of America (34%)
Reynolds Metals (28%)
Kaiser Aluminum and Chemical (24%)
Anaconda Aluminum (3%)
Consolidated Aluminum (1%)
Harvey Aluminum (3%)
Ormet Corp. (7%)

Copper
American Metal Climax, Inc. (7%)
American Smelting & Refining Co. (24%)
Anaconda Co. (20%)
Inspiration Consolidated Copper Co. (2%)
Kennecott Copper Corp. (21%)
Cerro Corp. (2%)
Phelps Dodge Refining Corp. (24%)

Copper Powders
American Metal Climax

Zinc
American Smelting & Refining (34%)
American Zinc Co. of Illinois (12%)
Anaconda Co. (35%)
Bunker Hill Co. (19%)

Lead
United States Smelting, Refining & Mining Co.

Magnesium
Dow Chemical

Tin (electrolytic)
Wah Chang Corp.

Tantalum
Fansteel Metallurgical
Union Carbide Metals

Manganese (electrolytic)
Foote Mineral Co.
Union Carbide
American Potash & Chemical Corp.

Manganese Dioxide (electrolytic)
Burgess Battery Co.
Bright Star Industries

National Carbon
Olin Industries
E. J. Lavino & Co.
American Potash & Chemical Co.

Sodium Metal
E. I. du Pont de Nemours & Co.
Ethyl Corp.
National Distillers & Chemical Corp.

Selenium
American Metal Climax
American Smelting & Refining
Kawecki Chemical
Anaconda Copper Co.
Kennecott Copper Corp.

Lithium
Foote Mineral
American Potash and Chemical
Lithium Corp. of America
Maywood Chemical Works
(Stepan Chemical)

Cadmium
American Smelting & Refining
Bunker Hill Co.
Anaconda Co.
American Zinc Co. of Illinois

Tin Plate
United States Steel (34%)
Bethlehem Steel Co. (16%)
National Steel (18%)
Jones & Laughlin (6%)
Wheeling Steel (4%)
Youngstown Sheet & Tube (7%)
Inland Steel (4%)
Republic Steel (3%)
Kaiser Steel (5%)
Granite City Steel (3%)

Gold
U.S. Mint
American Smelting and Refining
American Metal Climax
Anaconda Co.
Handy and Harman Co.
Engelhard Industries

Table 16 (continued)

Silver
American Smelting & Refining
American Metal Climax
Anaconda Co.
Handy & Harmann Co.
Engelhard Industries

Gold (by-product)
American Metal Climax
American Smelting & Refining
Anaconda Co.
Inspiration Consolidated Copper
Kennecott Copper Corp.
Cerro Corp.
Phelps Dodge Refining Corp.

Silver (by-product)
American Metal Climax
American Smelting & Refining
Anaconda Co.
Inspiration Consolidated Copper
Kennecott Copper Corp.
Cero Corp.
Phelps Dodge Refining Corp.
American Zinc Co. of Illinois
Bunker Hill Co.
United States Smelting Refining &
Mining

Platinum, Palladium, Nickel (by-product)
American Metal Climax, Inc.
American Smelting & Refining Co.
Anaconda Co.
Inspiration Consolidated Copper
Kennecott Copper Corp.
Cerro Corp.
Phelps Dodge Refining Corp.

Detinning
M & T Chemicals Inc.
National Steel Corp.
Vulcan Detinning Co.
The Chicago Detinning Co., Inc.

Electroplating—
automotive
General Motors
Ford
Many small job shops

Electrolytic Hydrogen
Air Reduction Co., Inc.

General Dynamics
(Liquid Carbonic Div.)
Burdett Oxygen Co.
Air Products, Inc.

Electrolytic Oxygen
Air Reduction, Inc.
General Dynamics
(Liquid Carbonic Div.)
Burdett Oxygen Co.
Air Products, Inc.

Hydrogen Peroxide (electrolytic)
Pennsalt Chemical

Sodium Chlorate
American Firstoline Corp.
American Potash & Chemical Corp.
J. T. Baker Chemical Co.
Harshaw Chemical Co.
Hooker Chemical Corp.
McKessons & Robbins, Inc.
Olin Mathieson
Pittsburgh Plate Glass Co.
Chipman Chemical Co., Inc.
J. F. Henry Chemical Co.

Potassium Chlorate
American Potash & Chemical Corp.
J. T. Baker Chemical Co.
Hooker Chemical Corp.
Mutchler Chemical Co., Inc.

Ammonium Perchlorate
American Firstoline Corp.
American Potash & Chemical Corp.
Orlex Dyes and Chemicals Corp.

Chlorine
Dow Chemical (30%)
Pittsburgh Plate Glass (14%)
Diamond Alkali (8%)
Allied Chemical (6%)
Olin Mathieson (6%)
Hooker Chemical (6%)
Wyandotte Chemicals (6%)
Pennsalt Chemicals (4%)
F.M.C. (3%)
Stauffer Chemical (3%)
Ethyl Corporation (3%)
E. I. du Pont de Nemours & Co. (2%)
Frontier Chemical (2%)

Table 16 (continued)

Monsanto Chemical (1%)
Jefferson Chemical (1%)
Kaiser Aluminum (1%)
National Distillers (1%)
Hercules Powder (0.4%)
Arkansas Louis. Chemical (1%)
Velsicol Chemical (1%)
Fields Paint Mfg. (0.1%)
General Aniline & Film (0.3%)
International Min. & Chem. (0.2%)

Sodium Hydroxide (caustic soda)
Dow Chemical (31%)
Pittsburgh Plate Glass (14%)
Allied Chemical (7%)
Olin Mathieson (7%)
Diamond Alkali (9%)
Hooker Chemical (6%)
Wyandotte Chemicals (7%)
Pennsalt Chemicals (5%)
FMC (3%)
Stauffer Chemical (3%)
Frontier Chemical (2%)
Monsanto (1%)
Kaiser Aluminum (1%)
Arkansas Louis. Chem. (1%)
Velsicol Chemical (1%)
Hercules Powder (0.3%)
General Aniline & Film (0.3%)
E. I. du Pont de Nemours & Co. (0.2%)
International Mining & Chem. (0.1%)
Fields Paint Mfg. (0.1%)

Storage Batteries
General Motors (18%)
   (Delco Remy Div.)
Electric Storage Battery (17%)
Eltra Corp. (Prestolite Div.) (8%)
Gould-National Batteries (15%)
Globe Union (12%)
Ford Motor Co. (5%)
   (Hardware & Accessories Div.)
General Battery and Ceramic (7%)
Other small producers (18%)

Primary Batteries (Dry & Wet)
Electric Storage Battery (25%)
Union Carbide (25%)
   (Consumer Products Division)

McGraw Edison (10%)
   (Primary Battery Div.)
Yardney Electric Co. (8%)
Burgess Battery (4%)
Bright Star Industries, Inc. (4%)
Marathon Battery Company (4%)
Eagle Picher Co. (5%)
Mallory (P.R.) (2%)
Other small producers (13%)

Capacitors
Sprague Electric
General Electric
P. R. Mallory
Aerovox
Cornell Dubilier
Union Carbide
Many other smaller producers

Water Treatment
U.S. Office of Saline Water
Local Demineralization Plants

Electrochemical Measuring and Controlling Devices
Leeds and Northrup Co.
Honeywell, Philadelphia Division
Beckman Instruments
Bendix
Eberbach Corp.
Fisher Scientific Corp.
Curtis Instruments
Bissett-Berman Corp.
E. H. Sargent
Coleman Instruments
Greiner Scientific
Heath Company
London Co.
Brinkmann Instruments Inc.
Buchler Instruments
Ionics Inc.
Many other smaller firms.

Zinc Coatings
Amercoat
Humble Oil
Carboline Company
Socony Paint Products Co.
   (Valdura Coatings Div.)

Cathodic Protection
Cathodic Protection Service

Table 16 (continued)

Corrosion Rectifying Co., Inc.	Hinchman Company
Corrosion Services, Inc.	A. J. Smith Engineering Co.
Ebasco Services, Inc.	Steele & Associates, Inc.
Harco Corp.	Electro Rust-Proofing Co.

V. GROWTH—RECENT EXPERIENCE AND FUTURE PROSPECTS

The growth experience of the electrochemical industry during the period 1958 to 1963 is analyzed in the first half of this section. The second half is devoted to a subjective analysis and some predictions regarding the future prospects of the electrochemical industry. The last section was prepared with the help of Professor John O'M. Bockris, director of the University of Pennsylvania's Electrochemical Laboratory.

The measure of growth which will be used in this section is the average annual percent change. The average annual percent change may be expressed as

$$\frac{X_{63} - X_{58}}{X_{58}}\left(\frac{1}{5}\right)$$

where X_{63} and X_{58} represent some measure of size in the years 1963 and 1958, respectively. The measure of size used will be either value added or shipments.

In order to measure the growth of *real* economic activity, the size measures used in the average annual percent change formula are expressed in *constant dollars*, i.e., adjusted for price changes. The price indices used to adjust the value added or shipments figures are the Bureau of Labor Statistics' wholesale price indices. If the reader is uncertain of the technical meaning and use of price indices, he is referred to page 261 of this paper, where the meaning and use of price indices is summarized.

1. Relative Growth

In Table 17, value added in constant dollars for 1958 and 1963 is used as the measure of size to determine the average annual rates of growth for the electrochemical industry, U.S. GNP, all U.S.

manufacturing, and the industrial chemical industry. The constant-dollar figures for 1958 and 1963 were obtained by dividing the current-dollar figures by a suitable price index. The base period for all price indices used was 1957–1959.

The wholesale price index which was constructed in section II for the electrochemical industry was used to express the 1958 and 1963 value added by the electrochemical industry in constant (1957–1959) dollars. In the case of all U.S. manufacturing, the Bureau of Labor Statistics all commodities except farm products and foods wholesale price index was used. The industrial chemical industry's reported dollar value added was reduced to constant dollars by using the Bureau of Labor Statistics industrial chemical wholesale price index.[17] The GNP for 1958 and 1963 in constant dollars was obtained from published data. Value added was used as the measure of size because data on total shipments for all U.S. manufacturing were not available in 1958.

The U.S. GNP figure for 1958 and 1963 is in billions of constant dollars. These GNP figures may be interpreted as the sum of the value added by producers of *all* goods and services in the U.S. economy in 1958 and 1963. It is noted that in Table 17, the value of all U.S. manufacturing is only about $\frac{1}{3}$ that of U.S. GNP. This is because manufacturing is only one of the many economic activities which "yield" value added to the national economy.

The average annual percent changes in Table 17 were derived from the constant dollar value added figures for 1958 and 1963, which are also given in the table. The average annual percent

Table 17

Average Annual Percent Change in Value Added in Constant Dollars for 1958–1963

Component	Value added (constant dollars) 1958	1963	Average annual percent change
U.S. Electrochemical industry	$2291.1 MM	$3124.2	+7.3%
U.S. GNP	$444.5 bl.	$546.8 bl.	+4.5%
All U.S. manufacturing	$141,980.2 MM	$188,674.3 MM	+6.6%
U.S. Industrial chemical industry	$6166.2 MM	$9529.5 MM	+10.9%

change in the *real* value added ("real" is synonymous with "price adjusted") by the electrochemical industry during 1958–1963 was +7.3%. This average annual rate of growth for the electrochemical industry was 2.8% larger than that for U.S. GNP, 0.7% larger than that for all U.S. manufacturing, and 3.6% smaller than that for the industrial chemical industry. All of the rates of change presented in Table 17 measure changes in the real economic activity of the given sector, since the effect of price changes has been removed.

2. Growth of Major Components

Table 18 shows how the seven major components of the electro-chemical industry fared in regard to growth. Shipments is used as the measure of size. The dollar value of shipments in each case was reduced to constant dollars through the use of the appropriate wholesale price index.* The average annual percent change in these price adjusted figures was then computed for the period 1958–1963.

The usually severe problems associated with using shipments data do not exist in the context of growth rates. In a growth rate, a measure of size for an economic activity at one time is compared with the *same* measure of size for the *same* activity at another time. Since the comparison is confined within a single activity and since value added–shipment ratios for a given activity do not usually vary with time, measures of growth derived from using shipments or value added data will not usually differ significantly.

The second fastest growing component of the electrochemical industry was the surface-finishing component, made up almost exclusively of electroplating. Surface finishing accounted for over 25% of the value added by the entire electrochemical industry in 1963. During the 1958–1963 period, the average annual percent change in surface finishing was +14.4%.

If tin plate had been included within the electroplating cate-gory, the rate of growth of the surface-finishing component would have been reduced, because of the very low growth rate registered

*In the case of surface finishing, water treatment, devices, and corrosion prevention, the all commodity except farm products and foods wholesale price index was used, since specific indices for these components do not exist. For nonferrous metals and industrial chemicals, the specific wholesale price indices for these categories were used. In the case of batteries and capacitors, the electrical machinery and equipment wholesale price index was used.

Table 18

Average Annual Percent Changes in Shipments (at Constant '57–'59 Dollars) of the Major Components of the Electrochemical Industry

Component	1958 shipments in constant dollars, $MM	1963 shipments in constant dollars, $MM	Average annual percent change
Electrowinning and refining of nonferrous metals (including by-products)	$3072.6	$3717.0	+4.2%
Surface finishing	650.4	1110.9	+14.4
Electrochemical synthesis	459.3	660.0	+8.7
Batteries and capacitors	548.7	809.3	+9.5
Water treatment	0.1	0.2	+20.0
Measuring and controlling devices	12.7	21.8	+14.3
Corrosion prevention	148.4	191.6	+5.8

by tin plate. The fastest growing component within the electrochemical industry was water treatment, whose value added increased from $.1 MM in 1959 to $.2 MM in 1963, equivalent to a 20% average annual percent change.

In Chart 9, the growth of the major components of the electrochemical industry is contrasted with the growth of U.S. GNP, all U.S. manufacturing, and the industrial chemical industry. All components of the electrochemical industry, except the electrowinning and refining component, grew at an average annual rate greater than that of total U.S. GNP. In Table 19, the reason for the low rate of growth for electrowinning and refining will become evident, viz. the small growth of tin plate during the period.

When compared with the average annual percent change for all U.S. manufacturing, five components of the electrochemical industry—water treatment, surface finishing, devices, batteries and capacitors, and electrochemical synthesis—registered faster rates of growth. Corrosion prevention and electrowinning and refining of nonferrous metals grew at a slower average annual rate than did all U.S. manufacturing. In percentage terms, approximately 48% of the entire electrochemical industry grew at an average annual rate less than that of all U.S. manufacturing and 52% at a greater rate.

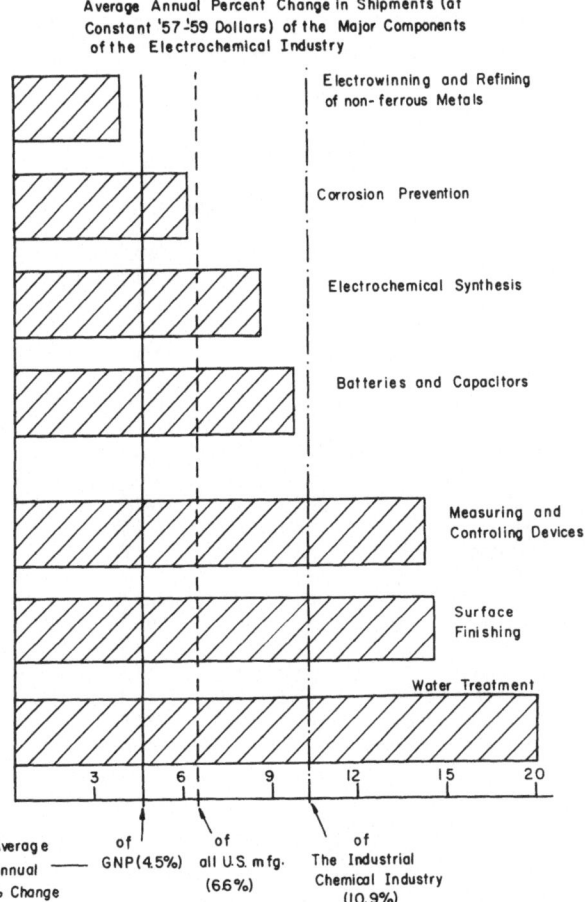

CHART 9

Average Annual Percent Change in Shipments (at
Constant '57-'59 Dollars) of the Major Components
of the Electrochemical Industry

Finally, in comparison with the industrial chemical industry, which grew at an average annual rate of 10.9%, over twice as fast as U.S. GNP, four components of the electrochemical industry representing approximately 73% of the industry, grew slower. The remaining 27% of the industry, represented by water treatment, surface finishing, and devices, grew at a faster rate than the industrial chemical industry. The electrochemical synthesis component

Table 19

Average Annual Percent Changes in Shipments (at Constant '57–'59 Dollars) of the Major Products in the Electrochemical Industry

Product	1958 shipments in constant dollars, $MM	1963 shipments in constant dollars, $MM	Average annual percent change
Aluminum	$837.7	$1080.7	+5.8%
Copper	989.3	1319.3	+6.7
Zinc	99.4	114.0	+2.9
Magnesium	33.8	55.2	+12.7
Tin plate	946.7	958.6	+0.3
Gold (excluding by-product)	43.7	33.2	−4.8
Silver (excluding by-product)	14.3	15.1	+1.1
Sodium	18.1	25.3	+8.0
Electroplating	643.8	1103.3	+14.3
Chlorine	199.1	316.8	+11.8
Sodium hydroxide	202.3	269.6	+6.6
Electrolytic oxygen	2.8	2.6	−1.4
Electrolytic hydrogen	6.8	10.6	+11.2
Hydrogen peroxide	22.5	5.9	−14.8
Storage batteries	360.8	489.5	+7.1
Primary batteries	135.4	195.3	+8.8
Capacitors (AC)	2.0	2.6	+6.0
Capacitors (aluminum)	35.2	64.1	+16.4
Capacitors (tantalum)	15.3	56.4	+55.7
Cathodic protection equipment	145.4	183.6	+5.3

of the electrochemical industry, which also accounts for a substantial share of the industrial chemical industry, grew at a rate 2.2% less than the industrial chemical industry.

3. Growth of Major Products

Table 19 gives the average annual percent changes for the major products of the electrochemical industry for the period 1958–1963. The fastest growing of the major products of the electrochemical industry was tantalum capacitors, which grew at an average rate of +55.7%. Aluminum electrolytic capacitors were second in terms of growth with an average annual rate of +16.4%. Water treatment by electrodialysis grew at an average rate of 20%. It, however, was not considered as a major product because of its small value added. Electroplating, the single most important product of the electrochemical industry, grew at an average annual rate of $+14.3\%$. Magnesium, fourth in terms of growth, grew at an average annual rate of $+12.7\%$ for the 1958–1963 period. Since tin plate accounted for over $\frac{1}{3}$ of the nonferrous metals component in 1963, its very low rate of growth was the major influence causing the low rate of growth for the nonferrous metals group.

To summarize the growth experience of the electrochemical industry during the period 1958–1963, the following may be listed:

1) The electrochemical industry grew at an average annual rate of 7.3%. This rate was 2.8% faster than U.S. GNP, 0.7% faster than all U.S. manufacturing, and 4.3% slower than the industrial chemical industry.

2) The fastest growing components of the electrochemical industry were water treatment, surface finishing, measuring and controlling devices, and batteries and capacitors, in that order.

3) The fastest growing major products of the electrochemical industry were tantalum capacitors, aluminum capacitors, electroplating, magnesium, and chlorine, in that order.

4. Future Prospects

In this final section, we turn from an objective analysis of the facts concerning the growth experience of the electrochemical industry to subjective considerations concerning the major prospects for growth.

At this time, the most important development for future growth within the electrochemical industry appears to be the fuel cell. Most of the large aerospace, electronics, and chemical firms are currently experimenting with these devices. The most obvious application of the fuel cell lies in the "electrochemical engine," a fuel cell electric motor combination. Very high efficiencies and the absence of noxious and polluting fumes are among the characteristics of the "electrochemical engine" which make industry sources optimistic over its chances of capturing a portion of the market now served by internal combustion engines and other electric motors.

The potential market is very large. It encompasses applications in all those areas currently served by electric motors and internal combustion engines. Industry sources feel that special purpose fuel cells, mainly for military uses, will be produced in large numbers near the end of this decade. In five or ten years broad commercial applications are foreseen in tractors, autos, and central power generating stations.[18]

Many firms are currently engaged in fuel cell research. Among the larger firms are: Aerojet General, Allis Chalmers, Chrysler, Dow Chemical, Electric Storage Battery, Esso, General Dynamics, General Electric, Leesona, Lockheed, RCA, Union Carbide, United Aircraft, and Westinghouse.

Desalination of brackish water in inland regions is an area of the electrochemical industry which should also experience significant growth. The recent droughts throughout the country have given new impetus to desalination projects. Electrodialysis is currently used by municipalities within the United States for the desalination of brackish water. A recent report published by Ionics Incorporated of Cambridge, Massachusetts, the major supplier of electrodialysis plants for municipal use, states:

> "More than 1000 small and medium sized communities in the U.S. and Canada now use water which contains in excess of 1000 parts per million (ppm) solids. Their aggregate population is 3 million people. These communities, and a still greater number of cities which now have limited supplies of fresh water represent a great potential."[19]

The Electrochemical Industry's share of what appears to many as a future billion dollar desalination industry is uncertain. However, since the value of desalination within the electrochemical industry in 1963 was only $.2 MM, there is reason for optimism concerning the growth of this segment of the industry.

The use of certain nonferrous metals, which are produced electrochemically, is expected to increase in the future. The applications of aluminum are continually expanding, usually at the expense of steel.

In a recent feature article in the Wall Street Journal, the following observations on the use of aluminum were made:

> "...the spectacular success of the aluminum pull-top, which has now replaced tin-plated steel lids on an estimated 60% of the nation's beer cans.... Right now aluminum is making fresh inroads in trucks, trailers and railroad cars. It's finding increasing auto usage in trim, wheels and brake drums, and is replacing steel in some electric power substations and lightweight transmission towers."[20]

It should be noted that some of the new applications of aluminum are made at the expense of other metals included within the electrochemical industry. The replacement of the tin-plated steel can be cans made of aluminum is an example of this.

Although the electrochemical synthesis of organic chemicals is not expected to approach that of inorganic chemicals, recent applications indicate significant growth in the future from the very low levels of activity existing in this area in the past. The electrochemical production of tetraethyl and tetramethyl lead by Nalco Chemical, the electrochemical preparation of dialdehyde starch by Miles Chemical, and Monsanto's large electrochemical facility for the production of adiponitrile were mentioned in section III as examples of recent activity in the area of electrochemical synthesis of organic chemicals.

The area of corrosion prevention promises future growth. Dr. John O'M. Bockris, director of the Electrochemistry Laboratory of the University of Pennsylvania, believes that the amount of expansion in this area is largely dependent upon the degree of knowledge available. In 1949, Dr. Herbert H. Uhlig, of the Massachusetts Institute of Technology, estimated the annual *direct* loss resulting from protection costs and replacement of corroded equipment in the U.S. at $5.5 billion.[21] Approximately 40% of this total represents replacement due to corrosion and 60% represents anticorrosion protection costs. These figures indicate the magnitude of the corrosion problem in the United States. The estimate of over $2.0 billion for annual replacement costs because of corrosion reflects a potential market which will be tapped as knowledge of corrosion prevention progresses. Although the figures cited above are for 1949, in a recent interview, Dr. Uhlig stated his belief that the figures are of the same magnitude for the current period.

The growth of electroplating will tend to follow the growth in automobile production. Since 1963, the terminal year for this study, significant growth has taken place in the number of automobiles produced. 1965 domestic automobile production will approach 9 million cars. This represents a 20% gain over the production of $7\frac{1}{2}$ million cars in 1963. Based on our estimate of $80 of plating per car, a nine million car year would raise the value of automotive plating to $720.0 MM from the level of $611.0 MM which prevailed in 1963. This represents an average annual rate of growth of nearly 10%.

There appear, therefore, to be many areas of potential growth within the electrochemical industry. It is difficult to judge which of these areas display the greatest potential. However, the potential for growth which exists within the electrochemical industry becomes clear if one imagines but $\frac{1}{5}$ of all current applications of internal combustion engines converted to "electrochemical engines" and but $\frac{1}{50}$ of all our drinking water derived from electrochemical desalination plants.

REFERENCES

[1] Definition taken in part from *A New Dictionary of Chemistry*, Ed. Miall and Miall, Longmans, Green, London, 1949.

[2] *1958 Census of Manufactures*, Vol. II, p. 11, U.S. Department of Commerce.

[3] *1958 Census of Manufactures*, Vol. I, p. 1–65, U.S. Department of Commerce.

[4] *1963 Census of Manufactures—Preliminary Report, Summary Series*, p. 4, U.S. Department of Commerce.

[5] Backman, *Chemicals in the National Economy*, Manufacturing Chemists' Association, Washington, D.C., 1964.

[6] *1958 Census of Manufactures*, Vol. I, p. 1–13, U.S. Department of Commerce.

[7] *1963 Census of Manufactures Preliminary Report, Summary Series*, p. 7, U.S. Department of Commerce.

[8] *Statistical Abstract of the United States, 1964*, p. 353, U.S. Department of Commerce.

[9] Backman, *Chemicals*, p. 46.

[10] *1960 Statistical Yearbook*, p. 33, International Tin Council, London, 1960.

[11] *1963 Minerals Yearbook*, pp. 555, 1011 and *1958 Minerals Yearbook*, pp. 484, 938, U.S. Bureau of Mines.

[12] *Industry Study*, National Assn. of Metal Finishers and Finishers' Management Magazine, Upper Montclair, New Jersey, 1965.

[13] *Plating* (April 1, 1950), p. 434.

[14] Ambler, *Economic Analysis of the Chlor-alkali Industry*, M.B.A. Thesis, University of Pennsylvania (1963).

[15] *Chem. Eng. News* **42** (1964) 52.

[16] Presented in brochure entitled "Sell Protection of Materials," prepared for *Materials Protection* Magazine, official monthly of the National Association of Corrosion Engineers.

[17] *1964 Statistical Abstract*, p. 351, U.S. Department of Commerce.

[18] *Industry Study Electronics-Electrical*, p. E21, a Standard & Poor's Publication (September 24, 1964).

[19] *Electrochemical Processes for Water, Food and Chemicals*, p. 7, Ionics Inc., Cambridge, Mass.

[20] *Wall Street Journal*, CLXVI (September 16, 1965), p. 1.

[21] *Chem. Eng. News* **27** (1949) 2764.

Index